DATE DUE

AP 23 '93			
FE 23 '96			

Demco, Inc. 38-293

D0488092

Some Recollections of Gap Jumping

Some Recollections of Gap Jumping

Sir Derek H. R. Barton

PROFILES, PATHWAYS, AND DREAMS
Autobiographies of Eminent Chemists

Jeffrey I. Seeman, Series Editor

American Chemical Society, Washington, DC 1991

Riverside Community College
Library
4800 Magnolia Avenue
Riverside, California 92506

QD22.B257 A3 1990
Barton, Derek, Sir, 1918-
Some recollections of gap
jumping

Library of Congress Cataloging-in-Publication Data
Barton, Derek, Sir, 1918–
    Some Recollections of Gap Jumping/Derek H.R. Barton.

    p. cm.—(Profiles, pathways, and dreams, ISSN 1047–8329)
    Includes bibliographical references (p.)and index.

    ISBN 0–8412–1770–X.—ISBN 0–8412–1796–3 (pbk.)
    1. Barton, Derek, Sir, 1918–   2. Chemists—
England—Biography.   3. Chemistry,   Organic—
History—20th century.   I. Title. II. Series.

QD22.B257A3   1990
540'.92—dc20
[B]   90–872
              CIP

The paper used in this publication meets the minimum requirements of American National Standard for Information Sciences—Permanence of Paper for Printed Library Materials, ANSI Z39.48–1984.

Copyright © 1991

American Chemical Society

All Rights Reserved. The copyright owner consents that reprographic copies may be made for personal or internal use or for the personal or internal use of specific clients. This consent is given on the condition, however, that the copier pay the stated per-copy fee through the Copyright Clearance Center, Inc., 27 Congress Street, Salem, MA 01970, for copying beyond that permitted by Sections 107 or 108 of the U.S. Copyright Law. This consent does not extend to copying or transmission by any means—graphic or electronic—for any other purpose, such as for general distribution, for advertising or promotional purposes, for creating a new collective work, for resale, or for information storage and retrieval systems. The copying fee is $0.75 per page. Please report your copying to the Copyright Clearance Center with this code: 1047–8329/91/$00.00+.75.

The citation of trade names and/or names of manufacturers in this publication is not to be construed as an endorsement or as approval by ACS of the commercial products or services referenced herein; nor should the mere reference herein to any drawing, specification, chemical process, or other data be regarded as a license or as a conveyance of any right or permission to the holder, reader, or any other person or corporation, to manufacture, reproduce, use, or sell any patented invention or copyrighted work that may in any way be related thereto. Registered names, trademarks, etc., used in this publication, even without specific indication thereof, are not to be considered unprotected by law.

PRINTED IN THE UNITED STATES OF AMERICA

1991 ACS Books Advisory Board

V. Dean Adams
Tennessee Technological
    University

Paul S. Anderson
Merck Sharp & Dohme
    Research Laboratories

Alexis T. Bell
University of California—Berkeley

Malcolm H. Chisholm
Indiana University

Natalie Foster
Lehigh University

Dennis W. Hess
University of California—Berkeley

Mary A. Kaiser
E. I. du Pont de Nemours and
    Company

Gretchen S. Kohl
Dow-Corning Corporation

Michael R. Ladisch
Purdue University

Bonnie Lawlor
Institute for Scientific Information

John L. Massingill
Dow Chemical Company

Robert McGorrin
Kraft General Foods

Julius J. Menn
Plant Sciences Institute,
    U.S. Department of Agriculture

Marshall Phillips
Office of Agricultural Biotechnology,
    U.S. Department of Agriculture

Daniel M. Quinn
University of Iowa

A. Truman Schwartz
Macalaster College

Stephen A. Szabo
Conoco Inc.

Robert A. Weiss
University of Connecticut

Foreword

In 1986, the ACS Books Department accepted for publication a collection of autobiographies of organic chemists, to be published in a single volume. However, the authors were much more prolific than the project's editor, Jeffrey I. Seeman, had anticipated, and under his guidance and encouragement, the project took on a life of its own. The original volume evolved into 22 volumes, and the first volume of Profiles, Pathways, and Dreams: Autobiographies of Eminent Chemists was published in 1990. Unlike the original volume, the series was structured to include chemical scientists in all specialties, not just organic chemistry. Our hope is that those who know the authors will be confirmed in their admiration for them, and that those who do not know them will find these eminent scientists a source of inspiration and encouragement, not only in any scientific endeavors, but also in life.

M. Joan Comstock
Head, Books Department
American Chemical Society

Contributors

We thank the following corporations and Herchel Smith for their generous financial support of the series Profiles, Pathways, and Dreams.

Akzo nv

Bachem Inc.

E. I. du Pont de Nemours
and Company

Duphar B.V.

Eisai Co., Ltd.

Fujisawa Pharmaceutical Co., Ltd.

Hoechst Celanese Corporation

Imperial Chemical Industries PLC

Kao Corporation

Mitsui Petrochemical Industries,
Ltd.

The NutraSweet Company

Organon International B.V.

Pergamon Press PLC

Pfizer Inc.

Philip Morris

Quest International

Sandoz Pharmaceuticals
Corporation

Sankyo Company, Ltd.

Schering–Plough Corporation

Shionogi Research Laboratories,
Shionogi & Co., Ltd.

Herchel Smith

Suntory Institute for Bioorganic
Research

Takasago International
Corporation

Takeda Chemical Industries, Ltd.

Unilever Research U.S., Inc.

Profiles, Pathways, and Dreams

Titles in This Series

About the Editor

JEFFREY I. SEEMAN received his B.S. with
high honors in 1967 from the Stevens
Institute of Technology in Hoboken, New
Jersey, and his Ph.D. in organic chemistry
in 1971 from the University of California,
Berkeley. Following a two-year staff fellowship at the Laboratory of Chemical
Physics of the National Institutes of
Health in Bethesda, Maryland, he joined
the Philip Morris Research Center in
Richmond, Virginia, where he is currently
a senior scientist and project leader. In
1983–1984, he enjoyed a sabbatical year at
the Dyson Perrins Laboratory in Oxford,

England, and claims to have visited more than 90% of the castles in
England, Wales, and Scotland.

Seeman's 80 published papers include research in the areas of photochemistry, nicotine and tobacco alkaloid chemistry and synthesis,
conformational analysis, pyrolysis chemistry, organotransition metal
chemistry, the use of cyclodextrins for chiral recognition, and
structure–activity relationships in olfaction. He was a plenary lecturer
at the Eighth IUPAC Conference on Physical Organic Chemistry held
in Tokyo in 1986 and has been an invited lecturer at numerous scientific meetings and universities. Currently, Seeman serves on the
Petroleum Research Fund Advisory Board. He continues to count
Nero Wolfe and Archie Goodwin among his best friends.

Contents

Photographs

Preface

"HOW DID YOU GET THE IDEA—and the good fortune—to convince 22 world-famous chemists to write their autobiographies?" This question has been asked of me, in these or similar words, frequently over the past several years. I hope to explain in this preface how the project came about, how the contributors were chosen, what the editorial ground rules were, what was the editorial context in which these scientists wrote their stories, and the answers to related issues. Furthermore, several authors specifically requested that the project's boundary conditions be known.

As I was preparing an article[1] for *Chemical Reviews* on the Curtin–Hammett principle, I became interested in the people who did the work and the human side of the scientific developments. I am a chemist, and I also have a deep appreciation of history, especially in the sense of individual accomplishments. Readers' responses to the historical section of that review encouraged me to take an active interest in the history of chemistry. The concept for Profiles, Pathways, and Dreams resulted from that interest.

My goal for Profiles was to document the development of modern organic chemistry by having individual chemists discuss their roles in this development. Authors were not chosen to represent my choice of the world's "best" organic chemists, as one might choose the "baseball all-star team of the century". Such an attempt would be foolish: Even the selection committees for the Nobel prizes do not make their decisions on such a premise.

The selection criteria were numerous. Each individual had to have made seminal contributions to organic chemistry over a multidecade career. (The average age of the authors is over 70!) Profiles would represent scientists born and professionally productive in different countries. (Chemistry in 13 countries is detailed.) Taken together, these individuals were to have conducted research in nearly all sub-specialties of organic chemistry. Invitations to contribute were based on solicited advice and on recommendations of chemists from five continents, including nearly all of the contributors. The final assemblage was selected entirely and exclusively by me. Not all who were invited chose to participate, and not all who should have been invited could be asked.

A very detailed four-page document was sent to the contributors, in which they were informed that the objectives of the series were

1. to delineate the overall scientific development of organic chemistry during the past 30–40 years, a period during which this field has dramatically changed and matured;

2. to describe the development of specific areas of organic chemistry; to highlight the crucial discoveries and to examine the impact they have had on the continuing development in the field;

3. to focus attention on the research of some of the seminal contributors to organic chemistry; to indicate how their research programs progressed over a 20–40-year period; and

4. to provide a documented source for individuals interested in the hows and whys of the development of modern organic chemistry.

One noted scientist explained his refusal to contribute a volume by saying, in part, that "it is extraordinarily difficult to write in good taste about oneself. Only if one can manage a humorous and light touch does it come off well. Naturally, I would like to place my work in what I consider its true scientific perspective, but . . ."

Each autobiography reflects the author's science, his lifestyle, and the style of his research. Naturally, the volumes are not uniform, although each author attempted to follow the guidelines. "To write in good taste" was not an objective of the series. On the contrary, the authors were specifically requested not to write a review article of their field, but to detail their own research accomplishments. To the extent that this instruction was followed and the result is not "in good taste", then these are criticisms that I, as editor, must bear, not the writer.

As in any project, I have a few regrets. It is truly sad that Egbert Havinga, who wrote one volume, and David Ginsburg, who translated another, died during the development of this project. There have been many rewards, some of which are documented in my personal account of this project, entitled "Extracting the Essence: Adventures of an Editor" published in CHEMTECH.[2]

Acknowledgments

I join the entire chemical community in offering each author unbounded thanks. I thank their families and their secretaries for their contributions. Furthermore, I thank numerous chemists for reading and reviewing the autobiographies, for lending photographs, for sharing information, and for providing each of the authors and me the encouragement to proceed in a project that was far more costly in time and energy than any of us had anticipated.

I thank my employer, Philip Morris USA, and J. Charles, R. N. Ferguson, K. Houghton, and W. F. Kuhn, for without their support Profiles, Pathways, and Dreams could not have been. I thank ACS Books, and in particular, Robin Giroux (acquisitions editor), Karen Schools Colson (production manager), Janet Dodd (senior editor), Joan Comstock (department head), and their staff for their hard work, dedication, and support. Each reader no doubt joins me in thanking 24 corporations and Herchel Smith for financial support for the project.

I thank my children, Jonathan and Brooke, for their patience and understanding; remarkably, I have been working on Profiles for more than half of their lives—probably the only half that they can remember! Finally, I again thank all those mentioned and especially my family, friends, colleagues, and the 22 authors for allowing me to share this experience with them.

JEFFREY I. SEEMAN
Philip Morris Research Center
Richmond, VA 23234

November 11, 1990

[1] Seeman, J. I. *Chem. Rev.* **1983**, *83*, 83–134.

[2] Seeman, J. I. *CHEMTECH* **1990**, *20*(2), 86–90.

Editor's Note

SIR DEREK BARTON was not born to the advantages and privileges his title might suggest. As he himself states on the opening page of this book, "Anyone who knew my family background would never have predicted that one day I would . . . receive the Nobel Prize." Born into a family of carpenters, his future seemed assured—he would enter his father's business and eventually operate it.

After two years of doing just that, Barton became restless and felt "there must be something more interesting in life." Thus, after a year of Technical College to hone his academic skills, Barton entered Imperial College, where he earned his bachelor's degree in science (1940) and his doctorate two years later. He then spent two years in Military Intelligence as an analytical chemist and one year in private industry as an organophosphorus chemist before accepting "the most junior position at Imperial College"—teaching practical inorganic chemistry. This inauspicious move presented no clue as to how it would come to affect his life, but the tide was turning—the revolution was coming!

In 1949 Barton went to Harvard as a visiting lecturer, sitting in for R. B. Woodward, who was taking his sabbatical year in residence. As described in other volumes in the *Profiles* series (William Johnson, Koji Nakanishi, and Jack Roberts), Cambridge, Massachusetts, was exploding with brilliance and seminal insight, and Derek Barton, swept up in the excitement, flourished and published.

While Harvard may have offered him a glimpse of what might be, he still had dues to pay to the British academic system. Upon his return from Harvard, Barton spent the next five years at Birkbeck College. But Barton and his chemistry became "red hot", and recognition was not to be denied. Following a brief stint at Glasgow, Barton took a professorship at Imperial, a post he held for 20 years (1957–1977).

Barton's career path is noteworthy for a number of reasons. First, he did not achieve a position of academic prominence until his late thirties. Yet, years earlier, and "in my spare time", Barton published his Nobel Prize study, *The Conformation of the Steroid Nucleus*, which revolutionized chemistry. Second, by the time he reached Birkbeck, he had put aside conformational analysis and was beginning to concentrate on what would become a lifelong interest: the discovery of new chemical reactions, especially those dealing with radicals.

Nearing Britain's mandatory retirement age, Barton moved to Gif-sur-Yvette as director of research at the Centre National de la Recherche Scientifique (CNRS). An Englishman managing one of France's most prestigious laboratories was a particularly noteworthy accomplishment. His decade in France was marked by his prodigious output of more than 200 publications that he authored or co-authored. But again mandatory retirement was threatening his career. "First it was 70 (years old), then 68, then 65. At that point I was 67, so they left me alone for one more year!"

At age 68, Derek Barton changed scene again, traveling across the Atlantic to Texas A&M University in College Station, Texas! Someone who has spent nearly 70 years in London and Paris does not move some 60 miles northwest of Houston for the cultural opportunities. Rather, the university's desire and drive to expand its involvement in chemistry appealed to Sir Derek and resulted in a synergistic relationship.

Interestingly, Barton mailed his original manuscript for this volume on his last day at the CNRS (January 8, 1987). "It does not cover all my scientific work because I am planning a full biography. No one will want the latter if everything is in your book. However, the present (material) covers the highlights up to work started in about 1960." When requested to update his biography to the present, he at first begged off, explaining that he was setting up his new lab at Texas A&M and "therefore, it is physically impossible to write any more . . . " A few weeks later, he called me to announce that he was writing again. He doubled his original output, then added the three Codas—a series of anecdotes and reminiscences of people and events that have touched his life. Raymond Lemieux, a long-time friend of Sir Derek, was asked to read the Coda to a Coda. He was thoroughly absorbed in the text, emitting an occasional "h-m-m-m", a chuckle, or a smile. When he finished, he looked up from the text and said, "I've known Derek for over 40 years, and I didn't know this about him."

The IUPAC Meeting, Zurich 1955. From left to right: Carl Djerassi, Michail Shemyakin, the author, R. B. Woodward, Alexander Todd, Vladimir Prelog, and Melvin Calvin. The last five members of the group all became Nobel Prize winners in the years following, Barton in 1969 at the age of 51. This assembly of remarkable scientists illustrates two phenomena found elsewhere in the Profiles series: first, that genius attracts genius. These same men appear in many of the other volumes together, alone, or with others whose names are legendary in the scientific community. Second, that despite the individuality, strength of personality, and competitiveness of each of these figures, they have forged close bonds and work together toward common goals.

As a young man, Barton had a reputation for being tough, probably because of his intense drive, his singularity of purpose, his individuality, and his determination where his work was concerned—and because he expected the same from those who worked with him. The drive and the energy are still there: How many people would leap from country to country, then continent to continent to avoid retirement? But the realization that not everyone can or will work at his level of intensity has softened the edges a bit. Loyal, supportive, and appreciative are how those who know him describe him. Sir Derek is also known not to mince words nor to waste them; like Sir John Cornforth he is known for his brevity (e.g., his caption for his photo on page 84).

Barton's novelty is constant; the arena just varies. Over his 40-year career, he has made seminal contributions to many areas of organic chemistry. He was recently honored with the 1989 ACS Award for Creative Work in Synthetic Organic Chemistry. About this award he wrote, "The award is for the years 1982–1987, so I am very pleased. Not yet too old!" . . . to jump the gap.

"Many of the things I have done which I think now were satisfying, have involved not chains of logic, but reasoning where there had been a gap in the chain of reasoning, where I have been able to jump that gap." Barton has recently jumped a few more gaps, one across the ocean into American and Texan chemistry, another into his eighth decade.

There are yet more gaps to be jumped.

JEFFREY I. SEEMAN
Philip Morris Research Center
Richmond, VA 23234

April 5, 1991

Some Recollections of Gap Jumping

Derek H. R. Barton

Early Work

ANYONE WHO KNEW MY FAMILY BACKGROUND would never have predicted that one day I would go to Stockholm to receive the Nobel Prize. My grandfather and my father were carpenters. My father founded a wood business that prospered and that permitted me to go to a good private school. I was born in 1918, and I had to leave school without any qualifications on the sudden death of my father in 1935. After two years of doing my share of manual labor in the wood business in the Thamesside town of Gravesend, more widely know for its association with Dickens, I felt that there must be something more interesting in life. I decided to go to the university.

I spent one year in a technical college to pass the necessary examinations, including the Imperial College entrance examination, completed a Bachelor of Science degree in chemistry in two years in 1940, and continued in organic chemistry for the Ph.D., which was completed in 1942.

For economic reasons, I had to live at home and go to London University, which was two hours of travel away. It was not clear which college, of the ten or so that offered chemistry, I should attend. I secured a copy of the university calendar and compared the colleges. The only difference that I saw was that the fees at Imperial College were 50% higher than those at the other colleges. My conclusion that Imperial College was, therefore, 50% better was, in fact, an underestimate!

Pyrolysis of Chlorinated Hydrocarbons

During the war years, the Ph.D. was obtained quickly and was preferably on a subject of national interest. My own work was

With a faithful friend—Rag (1926).

With my mother, while in the wood business (probably 1936).

mainly on the synthesis of vinyl chloride, because the professor of organic chemistry then (I. M. Heilbron, later Sir Ian Heilbron) was connected with the Distillers Company, which was prepared to offer a fellowship for this work. It was a financially attractive fellowship, and I accepted it.

A special laboratory was set aside, and I worked in collaboration with M. Mugdan, a refugee German chemist in his 60s. Mugdan had spent all his life in industry, and because of the mercury poisoning he had suffered while working on electrochemical processes, he was not in good health. He gave me valuable training as to how a practical chemist in industry attacks an industrially important problem. The work with Mugdan gave me useful experience in heterogeneous and homogeneous catalysis.

This early work[1] on the synthesis of vinyl chloride eventually led to a series of papers[2,3] on the pyrolysis of chlorinated hydrocarbons, research that was, indeed, physical chemistry. It is interesting to examine how the work began, the studies of mechanism that emerged from this work, and how the pyrolysis of chlorinated hydrocarbons eventually merged with the method of molecular rotation differences to produce a general stereochemical rule on *cis* elimination.

The gas-phase pyrolysis of 1,1-dichloroethane (1) and 1,2-dichloroethane (2) quantitatively yields vinyl chloride (3) and hydrogen chloride. For my Ph.D. work[1] from 1940 to 1942, I studied the pyrolysis of these two dichlorides in a glass tubular-flow reactor. At the beginning of the experiments, the reaction was fast because the glass surface had a catalytic effect. This heterogeneous reaction was soon suppressed when the surface was "poisoned", and the reaction became homogeneous.

For several months, the two dichloroethanes decomposed at nearly the same rate and with the same activation energy. One day, I observed with astonishment that the 1,2-dichloro compound was decomposing much faster than the 1,1-dichloro compound. However, the result was variable from day to day. After I thought about this observation for some time, I associated the enhanced pyrolysis rate with the fact that I had recently used a new technique for the purification of the starting materials. The earlier work was done with material fractionally distilled to a constant boiling point. The new technique involved the treatment of the dichlorides with aqueous potassium permanganate or aqueous chromic acid before drying and fractional distillation.

1,1-Dichloroethane always decomposed at the same rate, no matter how it was purified. 1,2-Dichloroethane, on the other hand, when purified by the oxidative treatment, gave varied rates of decomposition from day to day. An uncontrolled factor that catalyzed the reaction with 1,2-dichloroethane, but not that with 1,1-dichloroethane, was involved. This factor was a leak that let in variable amounts of air. A controlled amount of oxygen or chlorine had a spectacular effect on the decomposition of 1,2-dichloroethane but not on that of 1,1-dichloroethane. Clearly, the mechanisms of decomposition were different. Eventually, I realized that the inhibitor in commercial 1,2-dichloroethane was ethylene chlorohydrin, which forms an azeotrope with the dichloride. My careful fractional distillation had, therefore, served no purpose. It was the oxidative purification that removed the inhibitor.

Inhibitors are often associated with chain reactions. I postulated that the inhibited pyrolysis of 1,2-dichloroethane was a unimolecular reaction, like that of 1,1-dichloroethane (scheme 1). In contrast, the catalyzed reaction of 1,2-dichloroethane involved a chain reaction with the propagation steps as shown in scheme 2. 1,1-Dichloroethane did not decompose by the chain mechanism, because the preferred attack of $Cl^{\cdot}$ on CH_3–$CHCl_2$ in a putative chain reaction would give, by attack on the weakest (tertiary) C–H bond, the radical CH_3–$C^{\cdot}Cl_2$, which clearly could not eliminate a chlorine atom at the β position.

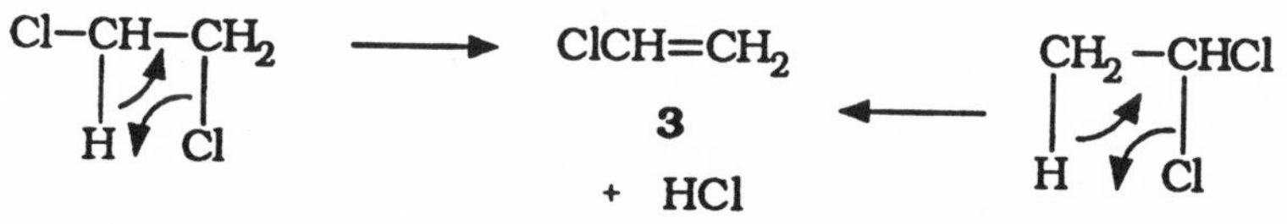

$$Cl\text{--}CH\text{--}CH_2 \longrightarrow ClCH{=}CH_2 \longleftarrow CH_2\text{--}CHCl$$

$$\textbf{3}$$
$$+\ HCl$$

$$\textbf{1} \qquad\qquad\qquad \textbf{2}$$

Scheme 1

$$Cl\cdot\ +\ CH_2Cl\text{--}CH_2Cl \longrightarrow {}^{\cdot}CHCl\text{--}CH_2Cl\ +\ HCl$$

$${}^{\cdot}CHCl\text{--}CH_2\text{--}Cl \longrightarrow CHCl{=}CH_2\ +\ Cl\cdot$$

Scheme 2

With more precise kinetics work,[2,3] these proposals were shown to be correct. The radical chain reaction often showed induction periods and was easily inhibited by olefins with an allylic hydrogen (like propylene) for facile reaction with a chlorine atom. 1,1-Dichloroethane and ethyl chloride showed only unimolecular decomposition, because the chain could not be propagated by the substrate.

As this work progressed, it became easy to predict the mechanism of decomposition of a chlorinated hydrocarbon. If the structure of the substrate gave a nonpropagating radical (as for 1,1-dichloroethane) or if the product of elimination had an allylic hydrogen (as for *tert*-butyl chloride and for the general family of chlorides), the radical chain could not be propagated, and the unimolecular mechanism was in effect.[4] The mechanism of decomposition could be predicted, and later work verified these predictions.[5]

A paper that I like to cite in discussions of mechanism is our study of the pyrolysis of 1,1,1-trichloroethane (4).[6] If the temperature and the ratio of surface area to volume are correctly chosen, then this substrate can decompose by three different mechanisms (reactions 1–3 in scheme 3) that proceed at the same rate. These observations should calm the spirits of those who want chemical mechanisms to be simple and unique.

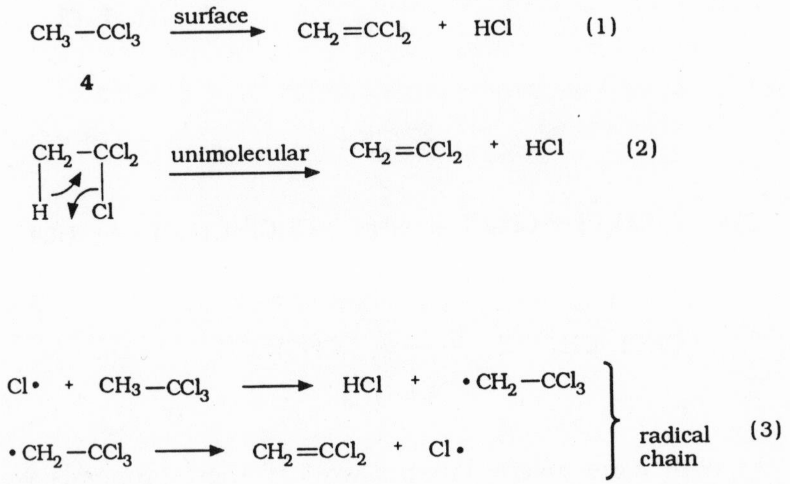

$$CH_3-CCl_3 \xrightarrow{\text{surface}} CH_2=CCl_2 \; + \; HCl \qquad (1)$$

4

$$\underset{\overset{|}{H} \; \overset{|}{Cl}}{CH_2-CCl_2} \xrightarrow{\text{unimolecular}} CH_2=CCl_2 \; + \; HCl \qquad (2)$$

$$Cl\cdot \; + \; CH_3-CCl_3 \longrightarrow HCl \; + \; \cdot CH_2-CCl_3$$

$$\cdot CH_2-CCl_3 \longrightarrow CH_2=CCl_2 \; + \; Cl\cdot \qquad \left.\begin{array}{c}\\\\\end{array}\right\} \begin{array}{l}\text{radical} \\ \text{chain}\end{array} \quad (3)$$

Scheme 3

Flour Beetles and Secret Ink

My first scientific paper was with Peter Alexander in 1943. It was sent to the *Biochemical Journal*, whose editor at first thought that the paper was a hoax. However, he accepted it.[7] When this work was started, Alexander was a research student in physical chemistry working on inert dust insecticides. He noticed that the flour beetle (*Tribolium castaneum*) produced a dark halo on the dust as it died. The beetle was clearly excreting a volatile chemical of some kind. I volunteered to identify the compound, or compounds, responsible for the effect.

Working during my spare time (evenings and Sunday; it was the beginning of my time with military intelligence), I grew 5000 beetles and collected them by negative geotropism. It sufficed to put a piece of paper above the flour onto which the beetles walked as the colony aged. When they were gently warmed in a flask in a current of air, the angry beetles gave off ethyl-*p*-quinone contaminated with some *p*-toluquinone. These substances were identified by collection in a dry-ice trap and reduction with sodium bisulfite to the corresponding quinols. *Tribolium* species produce quinones that react with the gluten in wheat flour to turn the flour pink, a typical reaction of the amino function with a *p*-quinone. I cite this story in some detail, because it shows that I thought then, as now, that chemistry is more interesting than spare time.

At the end of the Ph.D. work, I transferred into military intelligence, while still at Imperial College, and worked there for two years; the military intelligence work was mainly analytical chemistry, but it was also useful experience. We were engaged in the invention of secret inks. Although the message is not transmitted very quickly, the use of secret inks is a much safer procedure for the agent in the field than the use of a radio set. We took the spot tests beloved by analytical chemists and tried to turn them to advantage. Because an invisible message written in an aqueous fluid is very easily detected by an iodine spray even if pure water is used, either the sheet of paper has to be steamed after writing the message or one must use nonaqueous chemistry. We made considerable progress with inks based on nonaqueous chemistry.

Diversion into Industry

When it was clear that the war was coming to an end, we were encouraged to move into industry. I spent one year in Oldbury, Birmingham, working for Albright and Wilson, Ltd., on the synthesis of organophosphorus compounds. I did not invent the

Wittig reaction, or any other reaction, then, because I was too immature to know how to think at that time.

Everything started with electrochemically derived white phosphorus. At that time (1944–1945), we were not concerned yet with the effect of chemical operations on the environment. As one approached Oldbury, which was in a hollow, one could see a permanent fog produced by acid fumes. The trees had leaves for only a few weeks and then the leaves fell off. Albright and Wilson were not the only culprits. There was an old-fashioned lead chamber plant next door that made sulfuric acid. A canal ran through the plant, which was very convenient for the disposal of solid waste.

Fortunately, Albright and Wilson had an excellent library, and I continued to read extensively. At Imperial College, theoretical organic chemistry, as advocated by E. D. Hughes and C. K. Ingold, had not been highly regarded. In fact, we never heard about the work of Hughes and Ingold. It was a pleasure for me to begin reading about physical organic chemistry.

The industrial diversion was useful, although I did not find the work very challenging in its intellectual content. Industrial research today is very different, and the industrial laboratories are often more inventive than the academic ones. But I personally wanted to choose my own line of research, and again, I was convinced that there was something better in life.

On to the Academic Ladder

When I left industrial research, I was offered the most junior position at Imperial College (at half the salary I had been getting) to teach practical inorganic chemistry to mechanical engineers. I was delighted to accept, even though a more disagreeable task in academic chemistry would be difficult to find. After a month, I was invited to teach the professor's course on inorganic chemistry for engineers (of all types). Again, I jumped at the opportunity. After a year, I was promoted to teaching chemical kinetics to real chemists, which I did happily for three years.

At that time, H. V. A. Briscoe was professor of inorganic chemistry, and the other professor (of organic chemistry) was

Eighth Summer Seminar on the Chemistry of Natural Products, Grand Manan Island, Nova Scotia, 1956. First row, left to right: J. Fried, C. Djerassi, F. Anet, F. Toole, M. Kupchan, and H. Khorana. Behind them: H. Conroy, K. Wiesner, E. Wasserman, Z. Valenta, G. Stork, B. Belleau, the author, and B. Witkop. (Editor's note: According to Carl Djerassi, "We look like a group of criminals at a police lineup, with a few detectives—Khorana and Witkop—mixed in.")

I. M. Heilbron. My relations with Heilbron started out well, but I had the unfortunate tendency then—as now—to see the errors in other people's formulae. Sometimes there were errors in the Heilbronian formulae. Also, I had an equally bad habit of asking professors questions that they could not answer.

My relations with Heilbron were strained further when I left the organic chemistry department under a (slight) cloud and accepted the position in military intelligence associated indirectly with Briscoe rather than participate in the penicillin program that was being started in the organic chemistry department. However, when I moved into industry after two years with military intelligence, I accepted a position proposed to me by Heilbron and not an alternative (in the textile industry) suggested by Briscoe. So when I returned to Imperial College after a year with Albright and Wilson to work in Briscoe's department, the move did not find enthusiastic support from the organic chemistry department. However, the years passed, and in 1950, Sir Ian Heilbron helped me with his great influence to obtain the position at Birkbeck College and later put me up for the Royal Society. In the end, I was very grateful to both professors for their help in getting my foot onto the academic ladder and then helping me to climb it.

Method of Molecular Rotation Differences

An important example of my dedication to chemistry was the analysis of the literature of triterpenoids to correlate molecular rotation with structure. This work was suggested to me first by Sir Ewart Jones (at that time E. R. H. Jones).

Jones impressed all the students at Imperial College by his perfect lectures. I remember still the first lecture of his that I heard as a second-year student, which was on the aliphatic dicarboxylic acids, from oxalic acid and upwards. Later, Sir Ewart moved to the prestigious chair at Manchester University, and then in 1954, he succeeded Sir Robert Robinson at Oxford University. It would be difficult to find two men of such different temperaments as Sir Robert and Sir Ewart. The temperament of the latest Oxford professor of organic chemistry (Jack Baldwin) is much closer to that of Sir Robert.

Sir Ewart knew that I was an eccentric maverick, but he encouraged me all the same at critical moments in my life. Sir Ewart noted that a number of derivatives of lupeol (5) gave similar changes in molecular rotation upon acetylation, benzoylation, and oxidation to the ketone of the 3β-hydroxyl group. I

Sir Robert Robinson, ca. 1930.

E. R. H. Jones. (Photograph courtesy Vladimir Prelog)

5

surveyed all the available triterpenoid literature; certain regularities, indeed, could be observed, and some classification of compounds could be made.[8]

The method for these correlations is the principle of optical superposition, which was originally introduced in the last century by van't Hoff. With acylation reactions, new centers of chirality are not created, and we use the rule of shift.[9] Following

Jack Baldwin, relaxed and important. (Photograph courtesy C&E News)

the pattern of earlier work by E. S. Wallis and his colleagues,[10] I analyzed all published data on steroids.[11-13] The literature for steroids was much richer than that for the triterpenoids. Because I kept comparing differences in rotation, I called my approach "the method of molecular rotation differences".

Because the method has no real theoretical basis, the conclusions drawn must always be treated with reserve. The farther the UV absorption bands are from the line (usually the sodium D line) used for measurement, the more certain are the correlations. Unsaturated substituents interact, and the interaction (vicinal action) that gives deviations from the calculated $[M]_D$ (molecular rotation, sodium D line) data must decrease as the substituents are farther away from each other. Despite these reservations, the method was useful in the period 1940–1960 and can still be applied today.[14]

For example, stenols that contained one ethylenic linkage gave different molecular rotation differences (Δ values) for the changes from hydroxyl to acetate (Δ^1) and from hydroxyl to benzoate (Δ^2), depending on the position of the double bond. A saturated stenol (6) gave −34° and +2° for Δ^1 and Δ^2, respectively; a Δ^{14} stenol (7) gave −35° and +30°, a Δ^8 stenol (8) gave −40° and −42°, and a Δ^7 stenol (9) gave −6° and +30° for Δ^1 and Δ^2, respectively.

With these data, the structure of α-dihydroergosterol (5,6-dihydroergosterol), a classical compound to which structure 10 had been assigned, could be reconsidered. This compound gave −7° and +35° for Δ^1 and Δ^2, respectively, and clearly was of the 7(8)-double bond type (11) rather than the proposed 8(14)-double bond type (10). A critical evaluation of the literature[11a] showed that, indeed, structure 11 is correct; this structure has been reconfirmed many times since.

Encouraged by this first paper on the correlation of steroid structure and molecular rotation, I published a second paper on polyunsaturated sterols a year later.[11b] Again new structures could be proposed. For example, ergosterol D had been assigned structure 12, which was compatible with the older formulation (structure 10) for α-dihydroergosterol. Indeed, the reaction of α-dihydroergosterol with mercuric acetate was used to prepare ergosterol D. The new structure for α-dihydroergosterol, 9, suggested that ergosterol D must be 13. A critical reading of the older literature showed that there was already much evidence for the new structure 13.

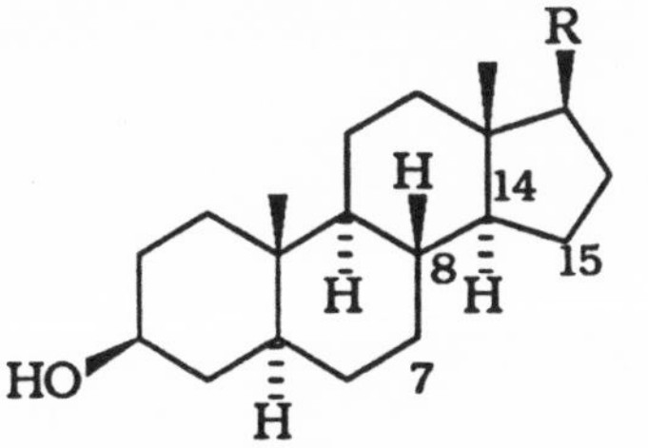

6

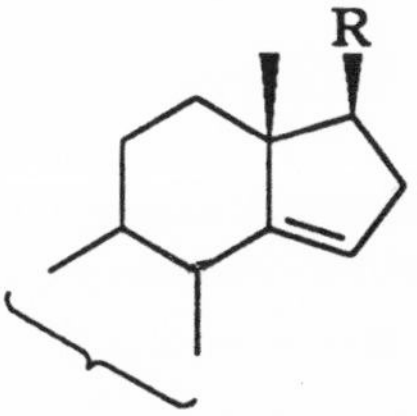

7

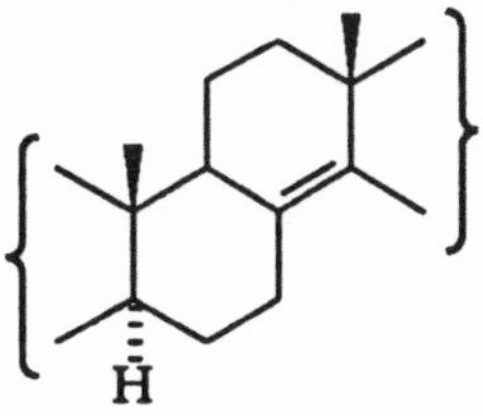

8

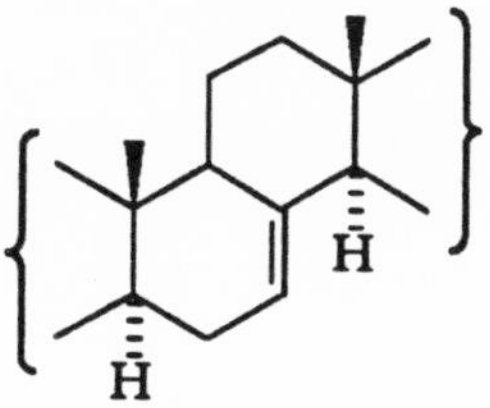

9

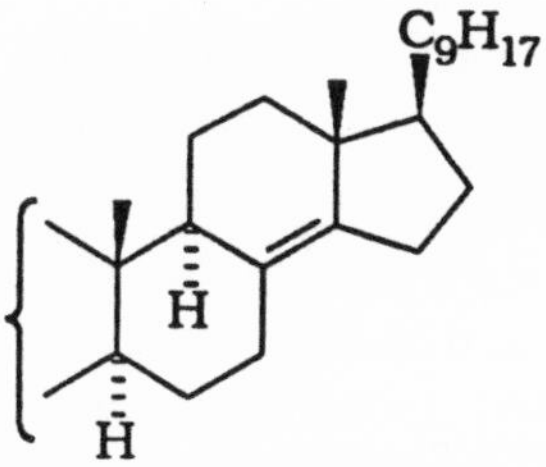

10

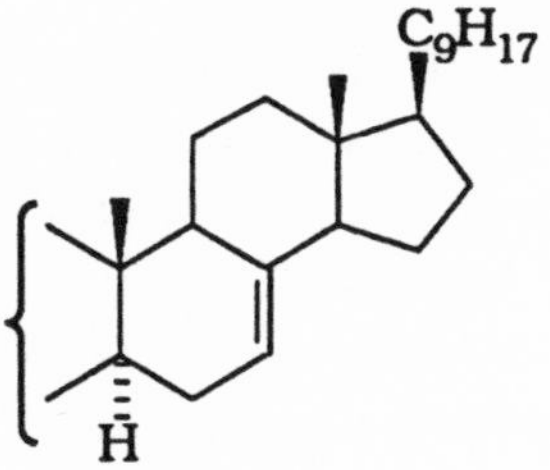

11

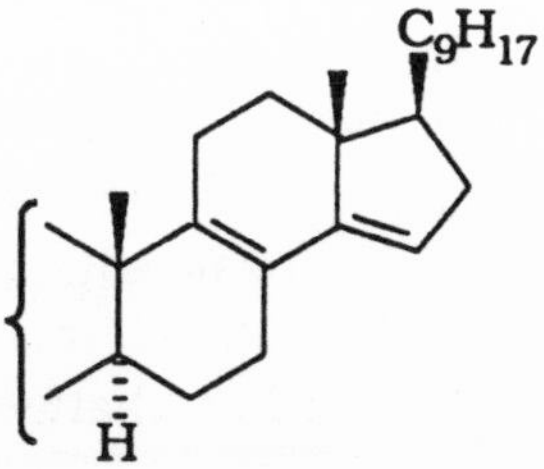

12

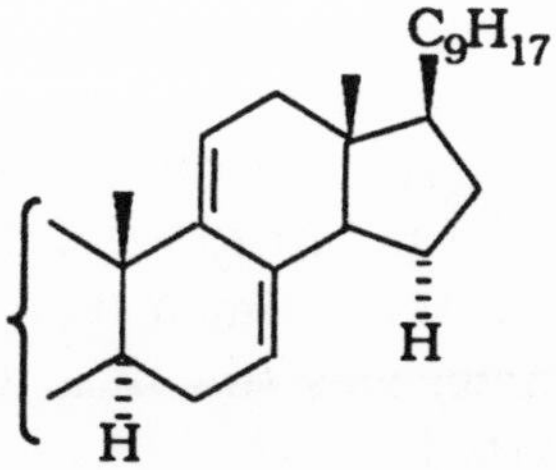

13

In the same year, I published a third paper[11c] on the correlation of molecular rotation with structure for hormones and bile acids. Although this paper was accepted without difficulty, the referees suggested that whereas correlation of published literature data was a good way to suggest new formulae, it would be even better to confirm these structures by some practical work. I could only agree, and I set to work, at first alone and then with a gifted young colleague, J. D. Cox. Cox thought that he was going to do physical chemistry with me, but he ended up making sterols. We published a number of papers confirming the earlier analysis of the literature.[12,13]

The simple examples of the method of molecular rotation differences might mislead the reader about the intellectual content of this subject. In fact, what was published was the integration of a large body of often confused and misleading literature. The interpretation required critical analysis and judgment. The method still finds application.[14]

The limits of the method of molecular rotation differences were also examined in relation to vicinal action,[13] and a summary paper was published in collaboration with (the late) W. Klyne.[15]

During our work on molecular rotation differences, I became familiar with the excellent papers by A. Windaus. It was natural to write and ask him if he could let us have a specimen of isodehydrocholesterol (14; cholesta-6,8-dien-3β-ol), which was a rare byproduct of the older Windaus procedure for the industrial conversion of cholesterol into 7-dehydrocholesterol (15; cholesta-5,7-dien-3β-ol). Windaus replied from Göttingen. After the war, he was still living in a part of his old house. His reply was not only courteous, but he also sent me his residual supply (about 50 g) of the diene. With this material we obtained, by hydrogenation, dihydrozymosterol (16; cholest-8-en-3β-ol), a derivative of the sterol zymosterol (17; cholesta-8,24-dien-3β-ol). After further correspondence, Windaus arranged a visit for me to a meeting of the German Chemical Society in Hanover. The paper I gave was published in *Zeitschrift für Angewandte Chemie*.[15] I had the honor to meet the great man and to speak with him in English. My German then, as now, was inadequate. Germany, at that time, was still in a terrible mess. How it has changed for the better!

14

15

16

17

Discovery of cis Elimination

Fate conspired to join our mechanistic studies on the pyrolysis of chlorinated hydrocarbons with our interest in the structure of unsaturated sterols and the method of molecular rotation differences. In 1948, Plattner and his collaborators[16] reported that the pyrolysis of the benzoate of 7-"β"-hydroxycholestanol acetate afforded pure cholest-7-en-3β-ol acetate. However, when we applied the method of molecular rotation differences to the so-called Δ^7-sterol, we found that the data were very different from those for authentic Δ^7-sterols (like α-dihydroergosterol, 11) and that the compound must have a different structure.

Plattner was an important figure at the Eidgenossische Technische Hochschule (ETH) in Zurich, and on the face of it, it would seem improbable that he could be wrong. I wrote to him pointing out that his sterol could not have a 7(8)-double bond. While on vacation in Switzerland in the summer of 1948, I went to Zurich for a day, and I was received in a friendly manner by Plattner. At first he wanted to send in a joint publication. However, this paper never arrived on my desk, and I finally wrote it myself. The critique appeared in due course in Helvetica Chimica Acta.[17] It was fortunate that my colleague W. J. Rosenfelder spoke German perfectly!

L. Ruzicka (photograph courtesy Vladimir Prelog) and Vlado Prelog.

It was perhaps unwise for a young man to criticize a distinguished professor at the prestigious ETH. There was worse to follow. In my 1950 article,[18] I showed that L. Ruzicka, also of the ETH, had made an error in the assignment of configuration at the C-3 position in ring A of triterpenoid alcohols. Ruzicka, one of the greatest organic chemists of his day, had received the Nobel Prize just before the war. He was a passionate and fiery man. Our relations for some years were confined to print and somewhat strained—at least on his part. However, he realized eventually that I had no intention to engage in polemics with him and that I had some talent for his own great passion— organic chemistry. I was finally invited to give a named lecture at the ETH in 1954, and we were friends from then onward. I know that he strongly supported my nomination for the Nobel Prize—along with that of Vlado Prelog.

What happened with Plattner's work was that the starting material was mainly the 7α-benzoate (18) and that the product of pyrolysis was the hitherto unknown Δ^6 isomer 19. We showed that the same compound (19) could be prepared by the pyrolysis of the 6β-benzoyloxycholestan-3β-yl acetate (20). In contrast, the pyrolysis of the 6α-benzoyloxy isomer (21) gave mainly cholesteryl acetate 22 (reaction 4).

When this work was carried out, the elimination reactions to form olefins were usually ionic *trans* eliminations. However, mechanistic considerations of the type developed for chlorinated hydrocarbons required that ester pyrolysis should be a unimolecular reaction with a *cis* elimination of the corresponding acid. The pyrolysis reactions of 18 and 20 to give 19 were clear examples of *cis* elimination in a unimolecular reaction.[19]

A diligent search of the literature revealed other examples of *cis* elimination, and a generalization could be made[20] that correlated this type of elimination with the unimolecular mechanism. A later study confirmed this mechanism for the pyrolysis of (−)-menthyl chloride, which showed first-order unimolecular pyrolysis giving *cis* elimination.[21] Thus the fusion of chemical kinetics and steroid chemistry gave us a generalization about *cis* elimination that still holds true to this day.

Grappling with Woodward

I first met R. B. Woodward in 1948 at Imperial College. Woodward was a precocious child and an equally precocious young man. I had just participated in the writing of a monograph on sesquiterpenoid chemistry,[22] and I was aware of some of the subtleties of the chemistry of α-santonin. Woodward gave a brilliant lecture on the structure of santonic acid. He spoke without notes or slides and covered the blackboard with beautifully drawn formulae. He argued that santonic acid, a compound obtained from α-santonin (23) during the classical work of the Italian school[22] and hitherto without a structure, was formed by base-catalyzed opening of the lactone ring with an $\alpha,\beta \rightarrow \beta,\gamma$ double-bond shift to give the ketoacid (24), the anion of which cyclized to give santonic acid, 25. He showed that santonic acid must be a derivative of cis-decalin.[23]

Every scientist must be judged by the standards of his time. In 1948, we had never heard anyone pose and then resolve a problem in such a clear and logical manner. The chemists of the period did not apply mechanistic thinking to problems in natural products chemistry. Woodward was the first to show us that problems in chemistry could be solved by thinking about

23

24   **25**

them. The curriculum at Imperial College at that time included only two one-hour lectures (in two years) in theoretical organic chemistry and never a question in the examinations. We were taught to think that mechanism had nothing to contribute to "real" chemistry. With one lecture, Woodward showed us the contrary.

I had long been corresponding with Louis F. Fieser of Harvard University about steroid chemistry. It was, therefore, not surprising that in the spring of 1949 I received a telephone call from Fieser inviting me to take Woodward's place for a year while Woodward was on sabbatical leave. I accepted at once. Bob Woodward's idea of sabbatical leave was to stay in his office and work harder than ever. This appointment at Harvard University gave me the chance to encounter his formidable brain at first hand for an extended period.

Gilbert Stork was also at Harvard at this time. Stork's imaginative intellect reinforced the exciting discussions about organic chemistry that invigorated the traditional Woodward weekly seminar. Usually a speaker would be recruited to talk

Louis Fieser in his Harvard laboratory during the 1940s. (Photograph courtesy Harvard University Archives)

about his work starting at (theoretically) 8 p.m., but in fact at approximately 8:30 p.m. The speaker would be closely questioned by all. At about 10 p.m., Woodward would pose a problem from the literature. Guests and students would spend up to an hour trying to solve this problem. When they had all failed, Woodward would give his solution, which was always correct. He would then call for anyone else to pose a problem. The problem being posed, it was always Woodward who proposed the correct answer first. However, this arrangement was somewhat unfair, because as midnight approached, Woodward became more and more brilliant, whereas I started to fall asleep, and even Gilbert Stork was relatively muted.

I was the first speaker of the season, and I talked about the problem that I had recently solved[19] of the double-bond position in the cholesterol benzoate pyrolysis product and how I had corrected the Plattner proposal.[16] At question time, I was astonished by the probing depth of Woodward's questions. I learned that I knew little compared with what he knew, but I learned quickly as the year went on. At least, he did not know as much chemical physics as I did!

Woodward's first major synthesis after quinine— performed with Doering—was that of the steroid nucleus, including cholesterol[24,25] and cortisone.[26] Woodward came to the United Kingdom in 1951 to deliver a Centenary lecture. He spoke about his brilliant total synthesis of steroids, and we were all impressed when he showed a formula (structure 26) and said that this compound is called chrismasterol, because it was first synthesized on Christmas day (1950). I can testify that I was in the Harvard Chemistry Department on the same day in 1949, and I was not alone.

26

The synthesis of cholesterol was also a subject of research for Robert Robinson for many years. Robinson's work was crowned by an elegant synthesis in collaboration with J. W. Cornforth.[28] By pure chance, the two great men met early on a Monday morning on an Oxford train station platform in 1951. Robinson politely asked Woodward what kind of research he was doing these days; Woodward replied that he thought that Robinson would be interested in his recent total synthesis of cholesterol. Robinson, incensed and shouting, "Why do you always steal my research topics?", hit Woodward with his umbrella. This story must be true, because Woodward told me about it several days later. Fortunately, after that incident, the two great men became firm friends, each recognizing the brilliance of the other. Robinson finally admitted that he had met someone who was his equal.

R. B. Woodward and W. von Eggers Doering celebrate the synthesis of quinine, in the mid-1940s. (Photograph courtesy Harvard University Archives)

It is unfortunate that Woodward is not alive to contribute to this series. However, knowing him as I did, I am sure that a contribution from him would have made the editor's life hell, and his book probably would never have been produced. Woodward was in every way an exceptional person. He had taught himself more chemistry by the age of 18 than the professors at Massachusetts Institute of Technology (MIT) had acquired in their lifetimes. Fortunately, the professors at MIT eventually recognized Woodward's astuteness in chemistry and allowed him to take the examinations without attending the classes.

When I first met Woodward, I immediately recognized my equal. My experiences at Harvard taught me to recognize his superiority. And yet, we were always different in our chemical interests, in our choice of problems, and in our way of thinking. Woodward had a formidable intellect to solve a problem by strict application of logic. I have attacked problems more by intuition, especially in the last two decades.

Over the years we were often very close. Without my wishing it, I was drawn into the complications of his family life. Because he did everything to extremes, I suppose that his early death was inevitable. I wrote a biography of Woodward covering the years up to the chlorophyll synthesis for the National Academy. This document has been censored, and portions of it now appear in this volume, but I hope that the unabridged version will appear one day. It was finished at Antibes on Christmas day 1984.

Sir Robert Robinson with R. B. Woodward, 1951. (Photograph courtesy J. D. Roberts)

Conformational Analysis and the Nobel Prize

It was during my year at Harvard, during the spring of 1950, that I wrote the short paper (four pages) for *Experientia*[18] for which I later shared the Nobel Prize with Odd Hassel. The paper was short because I had to type it myself! A detailed account of the history of this publication has appeared,[29] and what follows is a short resumé of the key points. An excellent book[30] on the history of conformational analysis is also available.

Hassel's Pioneering Work

I first became aware of Hassel's work through his article on the preferred conformations of *cis*- and *trans*-decalin in *Nature (London)* in 1946.[31] Hassel showed that *trans*-decalin (**27**) has a two-chair conformation (**28**) and that *cis*-decalin (**29**) also has a two-chair preferred conformation (**30**).

The two-chair preferred conformation for *cis*-decalin was surprising, because the organic chemistry textbooks all represented *cis*-decalin as having a preferred two-boat conformation (31). There was, of course, no significant evidence for 31 in the literature; it was simply that conformation 31 was easy to lay on a surface.

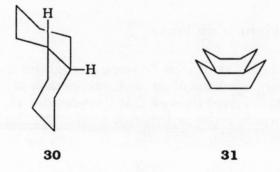

27   **28**

29

30   **31**

The Ethane Barrier

Even as a student, I was already aware of the problem of the ethane barrier. Until 1936, organic chemists considered that there was no barrier to rotation in ethane and related aliphatic molecules. It was a dramatic moment when Kemp and Pitzer[32] announced that the discrepancy between the entropy calculated by statistical mechanics and that observed experimentally for ethane can be explained by the existence of a rotational barrier of about 3 kcal. One of the two extreme conformations, the staggered (32) or the eclipsed (33), represented an energy minimum. Kemp and Pitzer[32] could not choose on the basis of the available evidence. If ethane had the staggered conformation, then cyclohexane would have a chair conformation. On the other hand, if the eclipsed conformation 33 were preferred, cyclohexane would have a boat, or even a planar, conformation.

32   **33**

Careful attention to the work of Hassel would have helped to solve the problem, but this attention to Hassel's work was not to be. The greatest theoretical chemist of the day, Henry Eyring, calculated the preferred conformation and reached the conclusion[33] that the eclipsed form (33) was the preferred conformation by about 3 kcal! The situation was complicated further by Langseth and Bak,[34] who studied cyclohexane by Raman spectroscopy and concluded that cyclohexane was planar. They also found that 1,1,2,2,-tetrachloroethane preferred the eclipsed conformation.

Hassel began his academic career at the University of Oslo in 1925. From the beginning, he was interested in structure determination by physical methods. He started with dipole

Henry Eyring at the Edgar Fahs Smith Lecture. Did we forget our 1939 calcu-lations? (Photograph courtesy C&E News)

moments and later added electron diffraction techniques. His careful work led him to the correct conclusions for almost all the cyclohexane derivatives with which he worked. However, his methods were not held in wide repute. Dipole moments were badly measured by some chemists (not Hassel), and at that time, electron diffraction was regarded as a subjective technique, because the darkness of the spots on the photographic plate had to be judged visually.

Hassel's work was not well appreciated. The situation was not helped by World War II and by the fact that Hassel's reply[35] to the article by Langseth and Bak[34] was published in Norwegian in an obscure journal in 1943. The publication in *Nature (London)*[31] was the first occasion that most chemists had to appreciate what Hassel had done.

First Applications of Conformational Analysis

When I read the publication by Bastiansen and Hassel,[31] I immediately set to work to see if the observed conformations could be calculated to be the most stable. These force field calculations were, in fact, the first in conformational analysis. The calculations were based on the equations of Hughes and Ingold[36] and of Westheimer and Mayer.[37] In each case, the calculated and the preferred conformations were in agreement.[38]

During my early work with conformational analysis, I was also interested in the stereochemistry of abietic acid (34). Oxidation of abietic acid with nitric acid catalyzed by vanadium pentoxide gives a relatively good yield of a tricarboxylic acid (structure 35). This acid was a *meso* compound (it has a plane of symmetry). What is the configuration at C-5? This problem was solved by pK_a measurements on the acid and its derivatives.[39]

At the Bürgenstock Conference in May 1965. The author is with Bastiansen, Eliel, and Sicher.

34

35

The analysis of the results was an example of conformational analysis, but because the question was relatively obscure and unimportant, nobody paid attention to this publication.

In 1948 I was already talking about conformational analysis and citing correctly the work of Hassel. However, I still had not seen the generality of the method. The paper that was finally sent to *Experientia* was a result of a seminar given at Harvard by Fieser. Fieser presented a theory about steric effects in steroids,[40] and I suggested that a much better approach would be to analyze the results in terms of the preferred conformation (36) of the molecule. It was very much to the credit of Fieser that he not only accepted my criticism but that he also suggested that I write my critique and send it to *Experientia* at the same time as his article. Because I thought that it was more likely for my article to be accepted this way, I did just as he suggested. Thus it was Fieser who inspired me to put on paper what I had been talking about for some time.

The article[18] rapidly changed the way that chemists thought about molecules. Indeed, in 1951, A. J. Birch[41] wrote, "Conformational Analysis for the study of the stability and reactivity of saturated or partly saturated cyclic systems promises to have the same degree of importance as the use of resonance in aromatic systems."

It is interesting to analyze the circumstances that permitted me, and not someone else, to write the *Experientia* article.[18] I acquired an excellent knowledge of terpenoid and steroid chem-

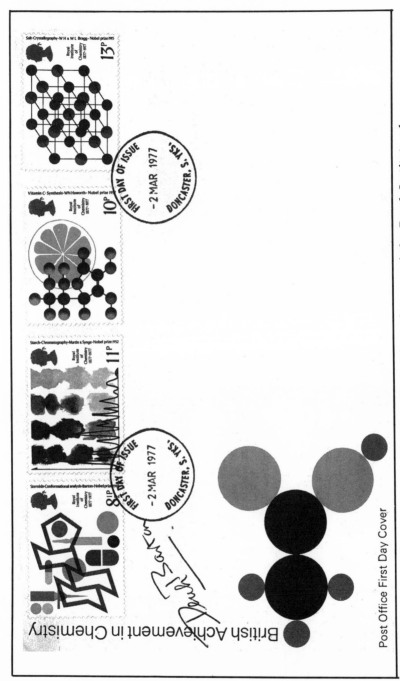

First-day cover celebrating the 100th anniversary of the Royal Institute of Chemistry. (Courtesy Laurence Harwood)

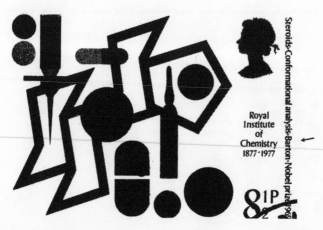

Close-up view of one of the stamps on the first day cover. Note Sir Derek Barton's name on the right-hand side of the stamp.

istry from my work with the method of molecular rotation differences. At the same time, teaching and researching in the field of gas-phase chemical kinetics brought me into contact with physical chemistry and fuelled my interest in mechanism. I saw that there should be a relation between rate data and preferred conformation for molecules derived from cholestane (**36**) and coprostane (**37**).

36

37

Arthur Birch when he was professor of organic chemistry, Manchester, England, mid-1950s.

Steroid chemistry was then already a mature and important subject,[42] and the discovery of the remarkable biological effects of cortisone in 1948 brought a large number of organic chemists to work in this area. The world was ready for formulae like **36** and **37**, which immediately correlated a large body of chemical evidence. There are eight secondary positions in the cyclohexane rings of **36** and **37**, and many pairs of hydroxyl isomers were already known. One could correlate their stability and reactivity, as well as those of their derivatives.[43]

Hassel or Pitzer could have written the 1950 article. But Hassel was a recluse and had no interest in chemical reactivity. Pitzer was then director of research of the Atomic Energy Authority and no longer engaged in research in the area. But, of course, I only anticipated what would have come in a few years anyway.

Disciples of Conformational Analysis

When the *Experientia* article[18] was still in press, I had the pleasure to make a short trip to the U.S. Midwest to visit the Universities of Notre Dame, Wisconsin, and Minneapolis. At Notre Dame, I talked about the conformation of the steroid nucleus, and E. L. Eliel, a young man like myself, immediately recognized

This picture was taken within a month or so of my writing the 1950 paper on conformational analysis. (Dupont Jackson Laboratory, Wilmington, DE)

the importance of the subject. Eliel eventually became the leading authority on the subject. His book,[44] *Conformational Analysis*, was published in 1965. The book was so well written that the American Chemical Society recently republished it in its original format. The most authoritative recent article, by J. I. Seeman,[45] contains a very detailed and clear study of the subject.

Despite my early interest in chemical kinetics, I never tried to study the quantitative aspects of conformational analysis. This subject was left to others.[44,45] I had had enough of chemical kinetics by 1950, and I thought that it would be more exciting, for me at least, to apply the principles of conformational analysis to important structural problems, which I did.

Conformational Analysis
in Structure Elucidation

Early in 1950, I was appointed reader (and later, professor) of organic chemistry at Birkbeck College, at the University of London. I was responsible for the teaching of organic chemistry, but I never forgot my experiences as a teacher of inorganic chemistry and physical chemistry. Birkbeck was (and is) the only college in the University of London that is a night school. Thus, one could carry out research all day and teach from 6 to 9 p.m. This system was excellent for research, but it was not appreciated very much by wives!

β-Amyrin

At Birkbeck College, I decided to apply the ideas of conformational analysis, a phrase first used in a publication by W. S. Johnson,[44,45] to stereochemical problems in triterpenoid molecules. The β-amyrin molecule (**38**) has eight asymmetric

At Birkbeck College, 1953. (Photograph courtesy J. D. Roberts)

With the Queen Mother during the opening of the new chemical laboratories at Birkbeck College, 1953.

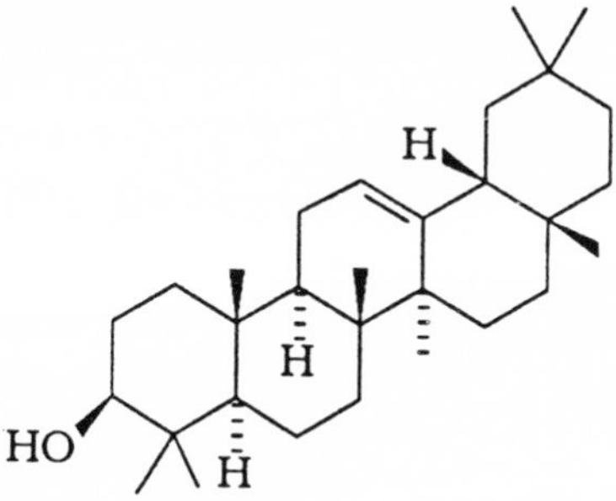

38

centers, and therefore, 256 isomers, made up of 128 pairs of enantiomers, are possible. The problem of the β-amyrin structure seemed very much more difficult than that of the steroid structure because of all the extra methyl groups. However, within a short time, the choice was narrowed down to one of two isomers,[43] and some timely X-ray work by Carlisle[46] put the finishing touch to the problem.

Lanosterol

Lanosterol (**39**) was another key molecule in structure analysis. As soon as its constitution was known, there were enough facts to write down the configuration of the molecule.[47]

  In 1954, Woodward and I collaborated on the total synthesis of lanosterol (by relay). The structure[47] of lanosterol (**39**)

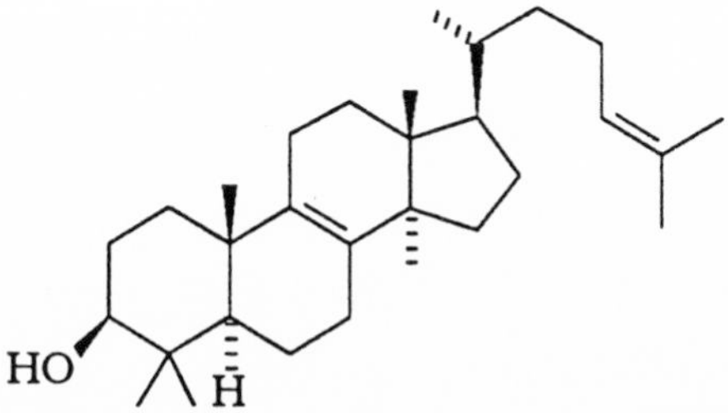

39

suggested to Woodward that the dihydro derivative of this biosynthetically important compound could be synthesized by the addition of three methyl groups to cholesterol, the total synthesis of which he had just accomplished. With a gifted collaborator (A. A. Patchett), Woodward devised a method to introduce the two methyl groups into C-4 of cholesterol. The step was unexpectedly easy. Methylation of cholestenone (**40**) with potassium *tert*-butoxide and methyl iodide gave the dimethyl derivative (**41**) in good yield (reaction 5). It remained to introduce the third methyl group[48,49] and to convert the saturated side chain of dihydrolanosterol into the 24(25)-unsaturated side chain of lanosterol (**39**). These steps were done at Birkbeck College with dihydrolanosterol as a relay.

With the aid of the information from Harvard, I wrote the full paper just before Christmas 1954 and sent it off to Woodward for his approval. I now mention a characteristic of the great man that has not been referred to before.[50] Woodward was always interested in the psychology of human behavior. In par-

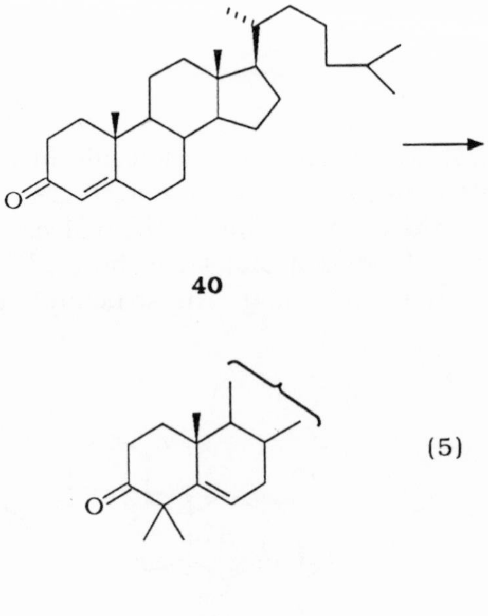

40

(5)

41

ticular, he liked to observe the reaction of someone to a mild stress imposed by himself. Woodward never acknowledged receipt of the full paper on lanosterol synthesis, although I was sure that he had received it. To play the game correctly, you had to assume the same indifference toward life as he pretended. I knew and I waited. Three years later,[51] he told me in casual conversation that he finally found time to read the manuscript and that it could be sent in to the *Journal of the Chemical Society* without the slightest change. Because I knew him well, I was not surprised. The event finally made us closer. He now had confidence in me, and I entered into the intimacy of his personal life.

At the Vitamin B_{12} Symposium in Zurich, 1975. Left to right: Lord and Lady Todd, R. B. Woodward, Dorothy Hodgkin, Vladimir Prelog, and T. Reichstein.

One of us is too thin! With Woodward (left) at the IUPAC Meeting in 1967.

Cycloartenone and Cycloartenol

During the analysis of the literature molecular rotation data for "artostenone", a supposed sterol derivative from the jackfruit, I was convinced that the formula proposed must be wrong. Eventually, I isolated again the beautifully crystalline compound and showed that it contained a cyclopropane ring and that it was a terpenoid.[52] Its structure was determined as the ketone corresponding to the alcohol **42**; suitable names that have been accepted for a long time were cycloartenone and cycloartenol for the ketone and the alcohol, respectively. Again, the structural work determined constitution and configuration at the same time. Lanosterol is a key compound in the biosynthesis of steroids in animals, and cycloartenol serves the same function in plants. Their biological importance was not suspected when the structural work was accomplished.

42

Sojourn in Glasgow

In 1955 I was appointed Regius professor of chemistry at Glasgow University, where I was made very welcome by Monteath Robertson. Robertson had been a graduate student at Glasgow University in the 1920s; he had worked on the structure of caryophyllene and other sesquiterpenoids. The methods of the time were hopelessly inadequate to solve such structural problems. In disgust, Robertson decided to become an X-ray crystallographer, which he did by working at the Royal Institution in London. For many decades he was a leading figure in X-ray crystallography, and it must have given him special pleasure to solve the caryophyllene structure by X-ray methods[53] at just about the same time as we did by chemical degradation.[54,55] Caryophyllene (**43**) was a remarkable structure for its time. It gave an astonishing series of tricyclic derivatives whose formation and stereochemistry were finally nicely explained by conformational analysis.[56]

Robertson and our group subsequently collaborated on a number of key structures. In each case, degradational chemistry and X-ray crystallography were started together. The most notable structures were clerodin[57] (**44**, the parent structure of the clerodanes), the very important limonin[58] (**45**, the parent of the

numerous limonoids), and glauconic acid[59] (**46**) and bysso-
chlamic acid[59] (**47**), members of the small, but original, nonadride
family. As computers improved, X-ray methods became largely
superior to chemical degradation, and after the nonadride work,
I participated in chemical degradation only in special cases.

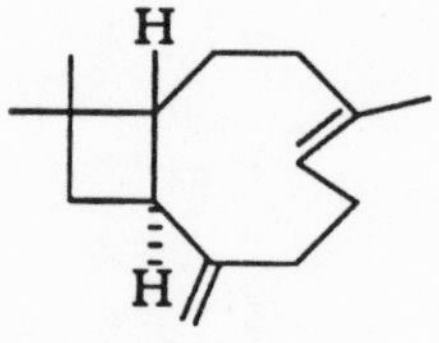

43

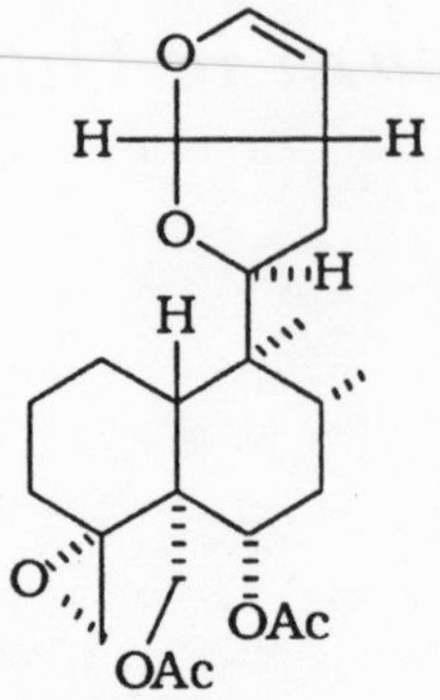

44

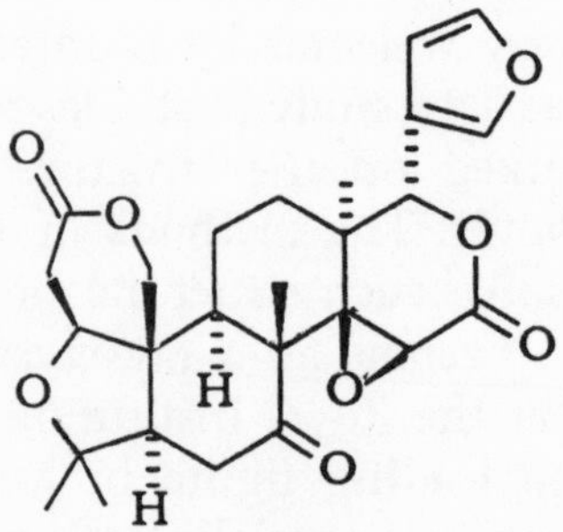

45

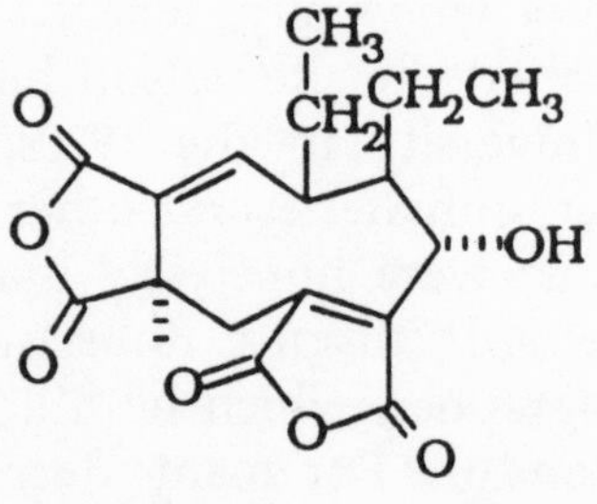

46

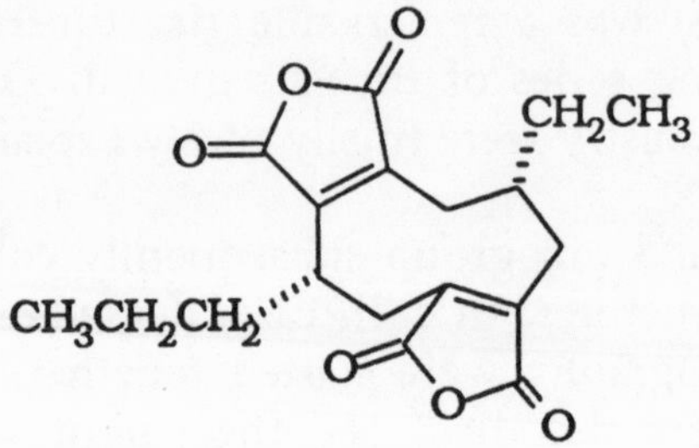

47

With Lord Todd, 1975.

It is worth explaining in some detail how I was appointed at Glasgow University and how I came to leave after only two years. It was known that Sir Robert Robinson was going to retire from Oxford University in 1954. I was foolish enough to think that I might secure this appointment, even though I was very young. At least one friendly voice (Lord Todd) advised me that I could not possibly be appointed at Oxford then and that E. R. H. Jones was destined to replace Robinson. Before the Oxford result was known, J. W. Cook of Glasgow became vice-chancellor at Exeter University, and Glasgow offered me the chair of chemistry. I temporized, and Glasgow said they would wait for a year. I agreed that if I was not appointed at Oxford, then I would go to Glasgow. When Jones was appointed in Oxford, I found myself on the way to Glasgow. Manchester University also expressed interest, but a bargain was a bargain. Worse, however, was the unexpected nomination of R. P. Lin-

Young and innocent—but not ignorant (Technion, Israel, 1955).

stead, professor of organic chemistry, to be rector at Imperial College. Eventually E. A. Braude was promoted from inside the department. Thus, I found myself in Glasgow.

All our chemical work on caryophyllene had been done at Birkbeck College. Clerodin and limonin began as structural problems in Glasgow and were continued in London after only two years in Glasgow. The work on nonadrides began in Glasgow and was continued in London.

Twenty Years at Imperial College

Pedagogic Innovations

Glasgow was the only place where all my requests for money were instantly accepted. However, I could not resist returning to Imperial College in 1957 when Braude committed suicide. Imperial College was not (and is not) the place for sensitive persons. It was nothing like Glasgow, and money, space, and students had to be fought for in a stern struggle. Nevertheless, I survived there for 20 years and managed to restructure the teaching of organic chemistry completely, with positive results.

Problem seminars and a tutorial system were progressively introduced in all years of the course. I learned these methods from Woodward. Examination questions now contained problems, as well as factual questions. One of our favorite problems for second-year undergraduates was the santonic acid structure disguised in various simplified forms. Amazingly, at least 20% of the class could do such problems without help. Had we then transformed 20% of them into young Woodwards? No, of course not. We simply started to teach students how to think better,

With R. B. Woodward and Ernest Guenther at the 1957 ACS meeting in Miami, where I received the Guenther Award.

and it was Woodward who initiated this process. Even if the student was perfect in the factual part of the questions, not more than 50% of the possible marks would be awarded for that part. At the same time, the marked emphasis on good practical work, especially in the third year, was maintained. By modern standards, all these reforms seem to be modest. But that life style of a chemistry student was new for the United Kingdom at that time and very different from the life I had known as a student 20 years before.

Further Research

Deviant Cyclohexane Boat. In 1957, McGhie[60] discovered the first example of a cyclohexane ring that had a choice of a chair or boat conformation and opted for the latter. Hassel surely was not pleased, although he never confirmed his displeasure to me. When dihydrolanosterone was brominated it gave two isomers as expected, the 2α-bromo ketone (48) and the 2β-bromo ketone (49). The 2α-bromo isomer (48) was more stable, with its bromine in the equatorial position, as shown by both infrared and ultraviolet spectroscopy. What was surprising was that the less stable 2β-bromo isomer (49) also had an equatorial C–Br bond, as determined by the same type of physical evidence. Clearly, compound 49 must exist in a boat-like conformation. Thus the 2α-bromo isomer had the normal chair conformation (50), and the 2β-bromo compound had an exceptional boat conformation (51). The importance of the sp^2 carbonyl group in the problem was shown by borohydride reduction, which smoothly gave the very hindered bromohydrin (52) with the C–Br bond in the axial position. These observations stimulated work that has continued to this day on the so-called 4,4-dimethyl effect. The original interpretation has been confirmed many times over.

Conformational Transmission. Eventually, conformational analysis became a routine tool. It was interesting to see if conformational effects were transmitted through molecules and for how far.[61] One started with dihydrolanosterone (**53**) and followed (by UV spectroscopy) the kinetics of its condensation with benzaldehyde under basic conditions to give benzylidene ketone (**54**). If the reaction rate for **53** is taken as 100, a shift of the double bond to the 7(8)-position (as in **55**) gave a fivefold reduction in rate. The rate for the saturated compound (**56**; relative rate of 44) was between those of **53** and **55**. These rate differences are major for a relatively remote nonpolar functional group (the ethylenic double bond). Because the origin of the effect is conformational, we coined the phrase "conformational transmission" to indicate the nature of this long-range effect.

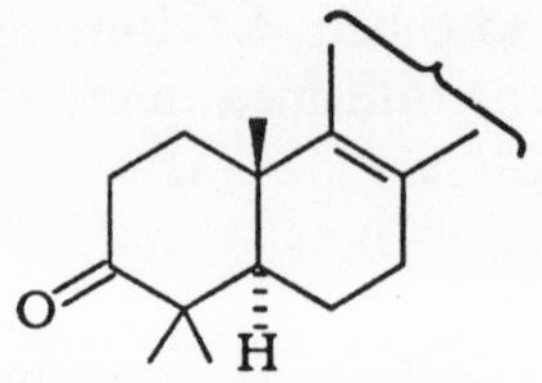

53

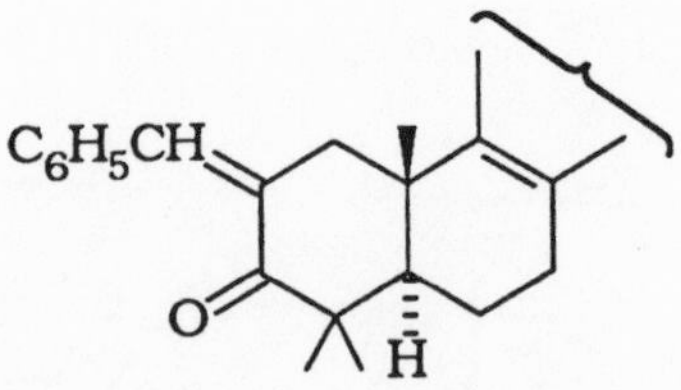

54

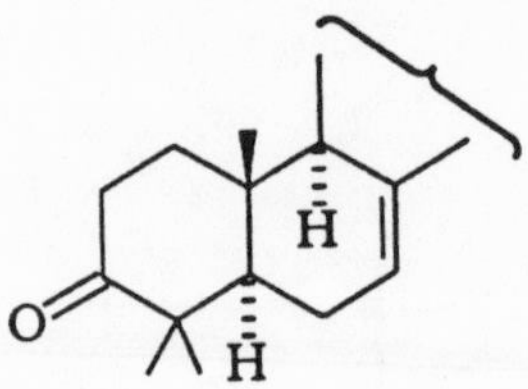

55

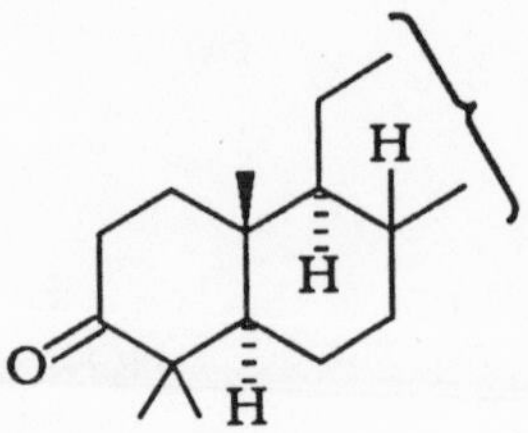

56

The long-range effect was seen even better in steroidal ketones (all cholestane derivatives). The reaction of the saturated structure **57** was four times faster than that of the 7(8)-olefin **58**. A shift in the double bond from the 7(8)-position to the 6(7)-position (as in **59**) increased the rate to 645, a 14-fold increase.[62]

57   **58**

59

More-remote functional groups also have distinct long-range effects, which in the case of the double bond or polyene function must be of a conformational origin. Polar groups can interact over the same distance by an electrostatic effect, as well as by conformational transmission and by orbital overlap effects. Allinger[63] correlated the postulated conformational transmission effects with force field calculations by assuming, as we had done, that the observed kinetics reflect the ease of formation of the 2(3)-enolate anion, which, after condensation with benzaldehyde and reenolization, eliminates hydroxyl to give the benzylidene ketone.

Many other examples of conformational transmission have been detected in steroids. The effect is seen in its most spectacular form in proteins, where the allosteric effect permits the interaction with a small molecule to change completely the conformation of the protein. Because the steroid nucleus is much less flexible than a protein structure, the magnitude of conformational transmission in steroids is remarkable.

Gap Jumping

Other directions for my research were opening up. I have sought always to see generalizations in chemistry—to see the relationship between facts that, to others, do not seem to be related. I have called this approach *gap jumping*.[64] In the conformational analysis story, one had to jump the gap between steroids and chemical physics. In succeeding endeavors, I jumped the gap between facts that were generally accepted to be true and the real truth, and then I saw the implications for biosynthesis.

One-Electron Oxidation

In a pioneering investigation, Pummerer[65] had discovered that the one-electron oxidation of *p*-cresol gave a nicely crystalline ketonic dimer. Structure **60** had been assigned on the basis that this dimer was a product of phenolate radical coupling and that treatment of **60** with acid gave a known diphenol **61** (scheme 4). Although structure **60** was widely accepted for three decades and was used even by Sir Robert Robinson and by C. Schöpf as a model for the biosynthesis of morphine and sinomenine, I felt

that it could not be true. Pummerer proposed that two phenolate radicals combined to give an intermediate **62** that then rear-
ranged to **60** (scheme 4). This rearrangement, in which a strong
aryl–hydrogen bond was broken, seemed to me to be kinetically

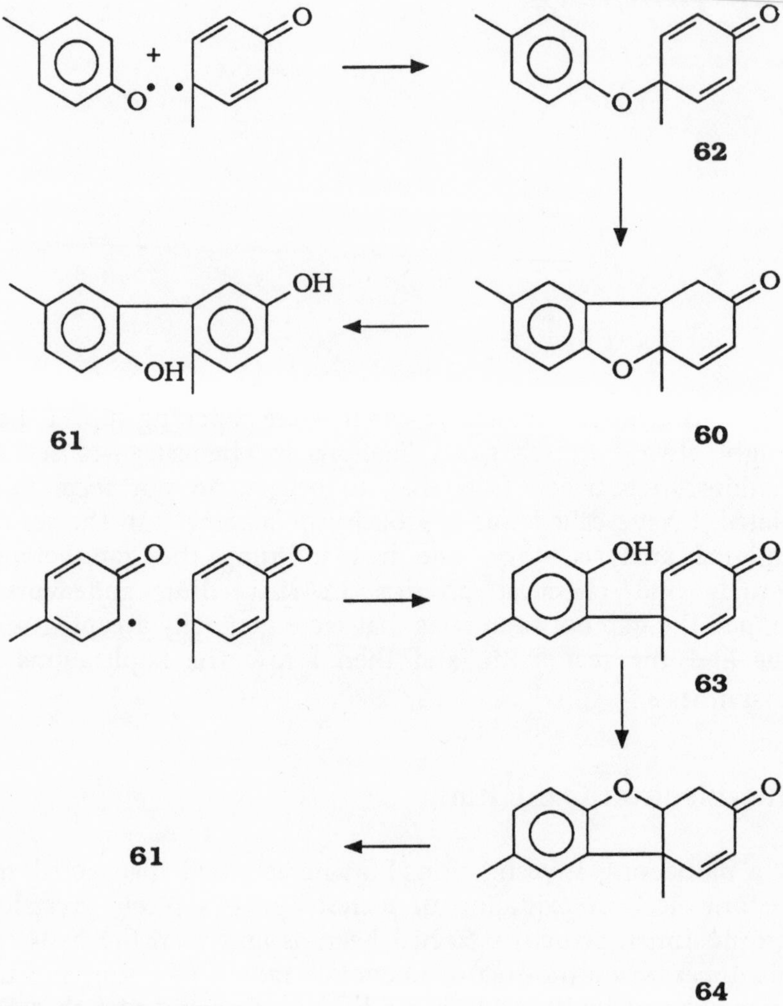

Scheme 4

With a friend, ca. 1961.

unacceptable. Morever, it was easy to write a different coupling of phenolate radicals to give **63**, which by simple Michael addition would afford **64** (scheme 4). Because **64** would give the diphenol **61** by an acid-catalyzed dienone–phenol rearrangement, I was convinced that I had the right formula. A new degradation of **64** gave 1-methylcyclohexane-1-carboxylic acid and proved that I was right.[66,67]

I at once realized that we had a model for a very simple synthesis of usnic acid **65**. The one-electron oxidation of methylphloracetophenone   (3′-methyl-2′,4′,6′-trihydroxyacetophenone) gave a dimeric product to which structure **66** was assigned. Brief treatment of **66** with cold sulfuric acid gave[66,67] racemic usnic acid, **65** (scheme 5). This synthesis was simple enough to satisfy my taste for elegance, although the overall yield was only modest (15%, unoptimized).

Biosynthesis of Phenolic Alkaloids

Because the wrong formula for Pummerer's ketone was used for previous ideas (*see* preceding section, One-Electron Oxidation) about the coupling of phenolate radicals (*meta* coupling) to give

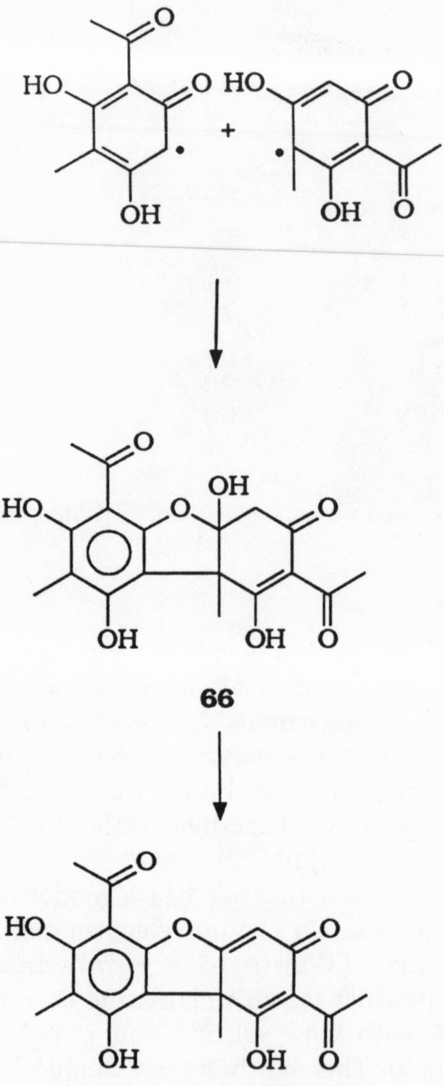

66

65

Scheme 5

morphine and sinomenine, these proposals must be in error. It was possible to write a new biogenetic hypothesis. I proposed that the benzylisoquinoline alkaloid **67**, then unknown but later found to be a common alkaloid (reticuline) in the plant kingdom, undergoes phenolate oxidation to give the dienone **68**, which, by ring closure, would give **69**, a strict analog of Pummerer's ketone. Reduction of **69** to an allylic alcohol and elimination of water would give thebaine (**70**). The steps from thebaine to codeinone (**71**) and then to morphine (**72**) were chemically plausible (scheme 6).

While I was in Glasgow, I was joined by a very intelligent young American colleague, Theodore (Ted) Cohen, now a distinguished professor in the University of Pittsburgh. He came to me, like many others, to work on conformational analysis. I was losing interest in this subject, and so we started to work on the biosynthesis of morphine instead of the deamination of aminocyclohexanes. Our practical aim was to synthesize reticuline, but after a few months, I was invited to write an article in a book dedicated to the great Swiss chemist, Arthur Stoll. This article was a good opportunity to summarize my views on phenolic coupling and biosynthesis. Ted Cohen did an excellent job of reviewing the literature, and the article that resulted has often been cited, especially because of the predictions it contained. People often protest that their work is not cited by others and

Ted Cohen in Glasgow, 1955—probably thinking about the coupling of phenolate radicals.

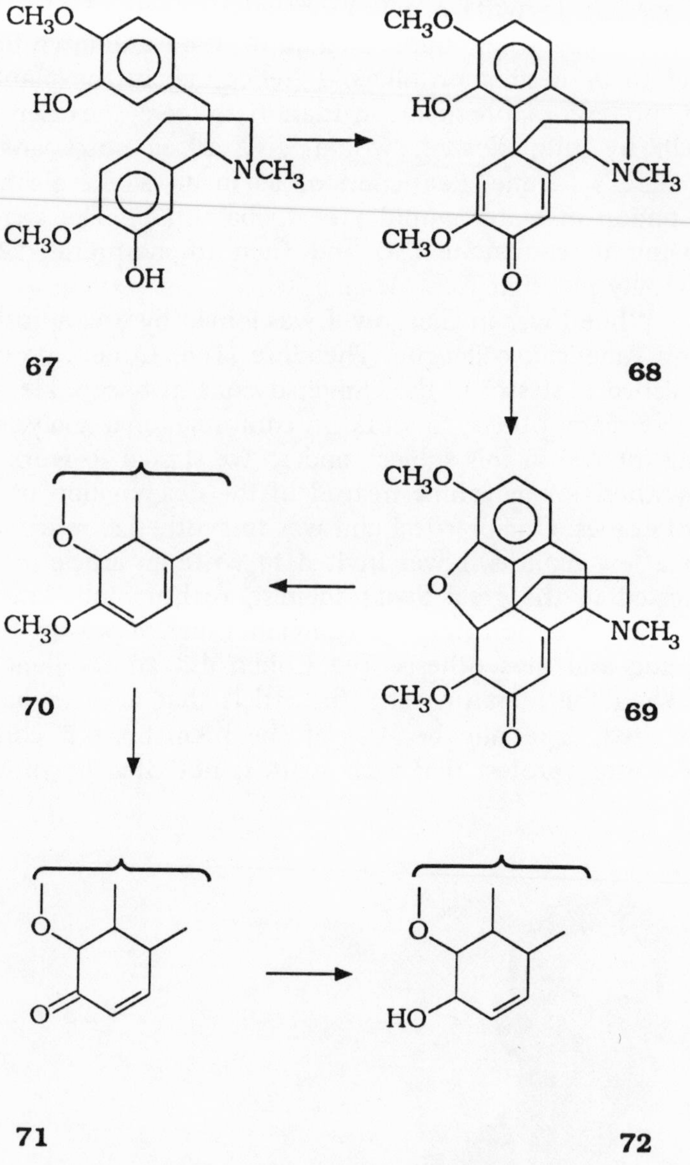
CH3O
HO
NCH3
CH3O
OH
67
CH3O
HO
NCH3
CH3O
O
68
CH3O
O
NCH3
CH3O
O
69
O
CH3O
70
O
O
71
O
HO
72

Scheme 6

claim that the fault is in their choice of a less well known journal. In fact, conformational analysis appeared in *Experientia* and the biosynthesis theory appeared in a book; in both cases, the ideas became widely recognized despite their humble origin.

A general article[68] based on a strict application of *ortho–para* phenolate radical coupling predicted the biosynthesis of morphine, as well as that of many other alkaloids. Even new types of alkaloids were proposed. It was very satisfying in later years to be associated with the structural work on crotonosine (73), which turned out to be an alkaloid of this new type.[69]

73

These biosynthetic proposals were tested by work with the radioactive tracers ^{14}C and ^{3}H, which became readily available at the right time. Considerable effort was devoted to the morphine alkaloids, and the overall correctness of the proposals was confirmed.[70] However, Nature does not strictly follow Pummerer. When the alkaloid 68 was synthesized[71] from thebaine, it did not spontaneously close to give 69. Later, 68 was found to be identical to a newly isolated natural product and was given the trivial name of salutaridine. Reduction of salutaridine (68) to the allylic alcohol 74 and very mild treatment with acid (scheme

7) gave thebaine (70) in good yield. The other steps (scheme 6; 70 → 71 → 72) were all confirmed in due course. Very small amounts of morphine could be chemically synthesized by oxidation of resolved reticuline (67) to salutaridine (68) and continuing the biosynthetic sequence to the end.

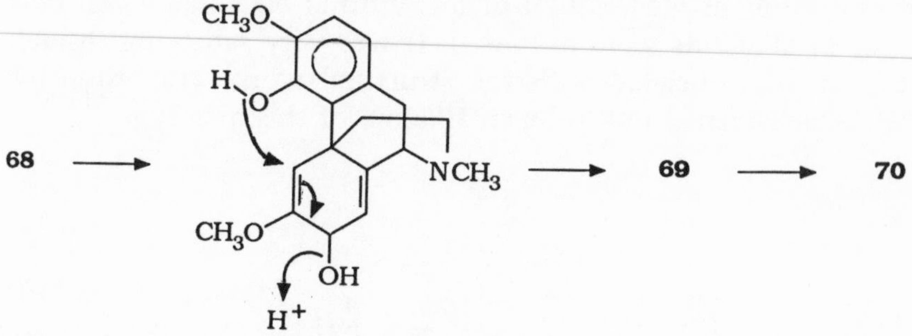

Scheme 7

Our work on the biosynthesis of phenolic alkaloids continued for nearly two decades. The *Erythrina* alkaloids were particularly interesting; the last paper[72] on this subject was in 1974. One day this theme will be developed elsewhere. It was, however, very satisfying to see how much elegant work (by Nature) came out of my original disbelief in the long-accepted formula of Pummerer's ketone.

Photochemical Reactions

I was involved in the discovery and invention of new photochemical reactions. My participation in this field started with a misconception.

In 1954 it was desirable to confirm the stereochemistry of the triterpenoid alcohol euphol (75) in terms of that of lanosterol (39) by an inversion of the C-10 methyl group in ring A. We knew how to transform 39 into the dienone 76. I thought that photolysis of 76 would break reversibly the weakest bond between C-1 and C-10 to give a familiar phenolate radical and a carbon radical, and that reformation of the C-1–C-10 bond would then set up a photochemical equilibrium between the stereoisomers **76** and **77**. A preliminary experiment[73] showed that something interesting happened in good yield, but the product was not the desired stereoisomer; the product had a new umbellulone chromophore.

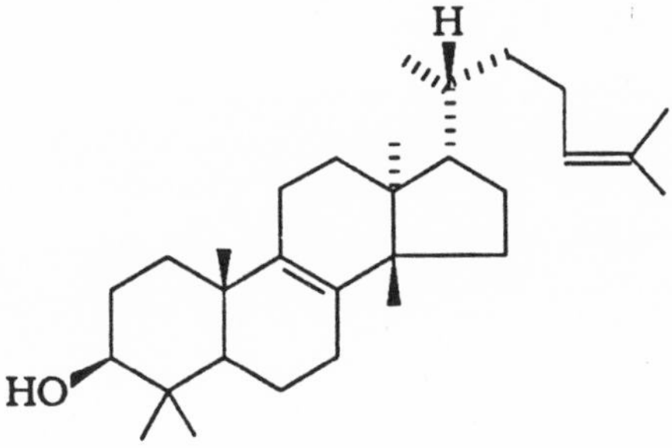

75

Because the dienone **76** required a multistep preparation, it was preferable to see what was happening to **76** by looking at what happened to the readily available α-santonin (**23**). Starting in 1955, we quickly determined[74] the structure of isophotosantonic lactone (**78**), a compound already prepared in the last century by photolysis of α-santonin in aqueous acetic acid (scheme 8).

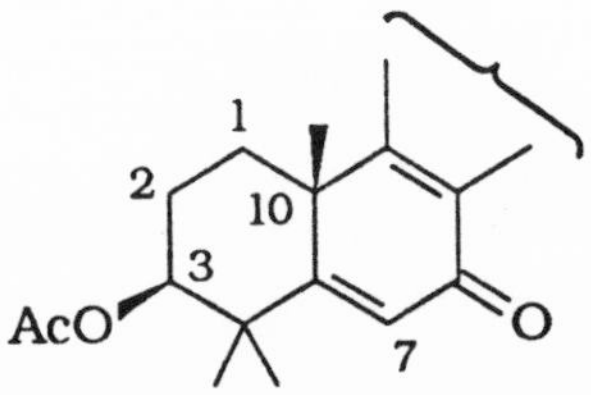

76

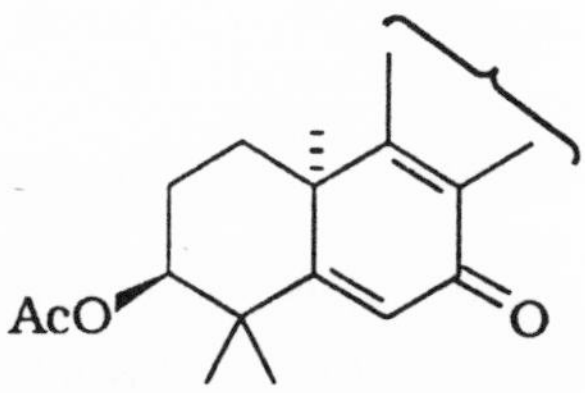

77

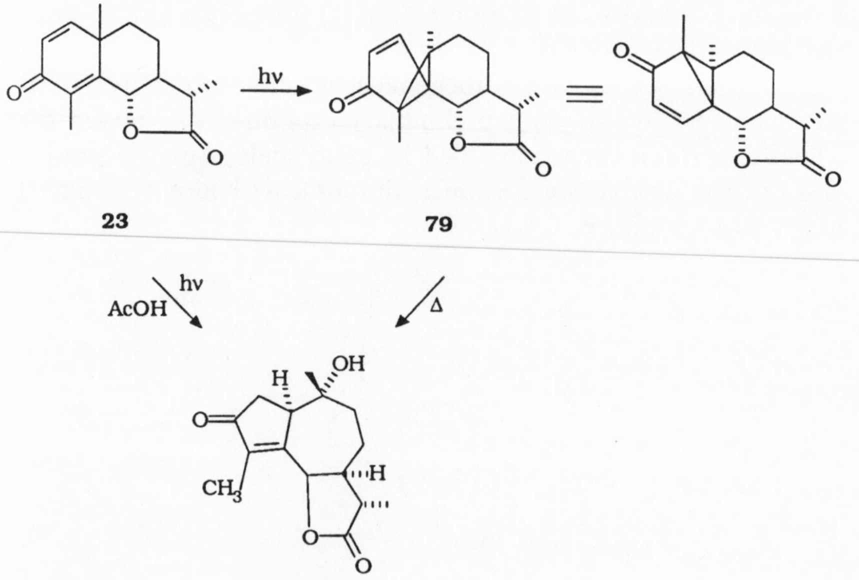

23   **79**

78

Scheme 8

Photolysis of α-santonin in neutral medium gave a new photoisomer, lumisantonin (**79**; scheme 8). Lumisantonin was also an umbellulone, as judged by its UV spectrum. On treatment with hot aqueous acetic acid, lumisantonin also afforded isophotosantonic acid (**78**; scheme 8). We[75,76] elucidated the structure of lumisantonin (**79**) at the same time as other colleagues[77,78] did. I believe that the remarkable molecular acrobatics of α-santonin in the presence of light had an important part in rekindling interest in organic photochemistry. Certainly, P. deMayo, who played a major role in this study, was encouraged to become a world expert on photochemical reactions in Canada in later years. It is amusing that my interest in photochemistry began with a misconception of what the effect of light should be on a cross-conjugated cyclohexadienone.

Later, we[79] showed that the rearrangement of **76** (reaction 6), which started off our interest in organic photochemistry, gave **80**, a product comparable with lumisantonin (**79**). This photochemical reaction is a general one. A thorough mechanistic study of this and related photochemical reactions has been carried out over several decades by H. E. Zimmerman (University of Wisconsin). The reaction has gone from an obscure observation of the 19th-century work on santonin to being one of the best investigated of all chemical reactions.

After the work on santonin, it was logical to examine the photochemistry of linearly conjugated dienones of type **81**. Fortunately, a very gifted colleague, G. Quinkert, now of Frankfurt, came to work with me at Imperial College. Our first experiments were promising. We showed that photolysis of dienones of type **81** set up a photostationary equilibrium with diene—ketenes of type **82**. In the presence of appropriate nucleophiles, these ketenes were quantitatively captured to afford[80,81] derivatives of type **83** (scheme 9).

Quinkert continued this work and related studies on organic photochemistry on his return to Germany. He became a world expert on the mechanism of photochemical reactions. It

Early photochemistry, ca. 1956.

81 ⇌ **82** →(NuH)→ **83**

Scheme 9

gave me special pleasure when a project that we had begun together in 1959 was finally completed[82] with exemplary attention to experimental detail. This project was an elegant, but practical, synthesis of dimethylcrocetin **84**. The photochemical cleavage of the bis(dienone) **85** (or a substituted derivative) was the critical step, with methanol acting as the nucleophile. The photochemical cleavage of dienones of type **81** is one of the best organic photochemical reactions that are known from the point of view of high chemical and quantum yields.

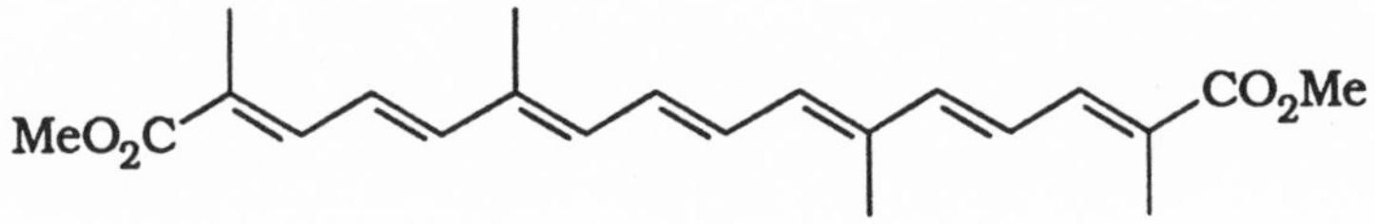

84

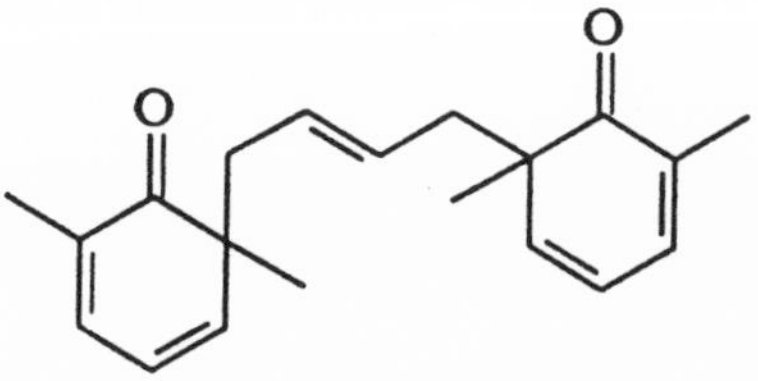

85

Synthesis for Medicine

In 1958, the Research Institute for Medicine and Chemistry (RIMAC) opened its doors in Amherst Street, Cambridge. It was originally a small brick-and-wood factory used for making cough mixtures. When refurbished, the laboratories were adequate for high-class biological and chemical research. The director of the institute was Dr. Maurice M. Pechet, a distinguished medical researcher and a world expert in steroidal hormones. He also had a Ph.D. in chemistry under Louis Fieser at Harvard. RIMAC was founded by the Schering Corporation to do original work in chemistry that could be applied to medicine. He asked me to accept responsibility for chemical synthesis. Dr. Pechet set the first objective as the synthesis of 17α-hydroxyaldosterone acetate (**86**). We eventually made this compound,[83] although we started off more modestly to seek a convenient synthesis of aldosterone acetate (**87**) itself.

As has happened on a number of occasions in my life, a good problem, a good idea, and a good experimentalist all came together at the same time. In 1958, there was an urgent need for a good synthesis of aldosterone. This vital hormone, which controls electrolyte balance in the body, was only available in trace amounts from natural sources. The hormone has a masked

Mine's bigger and better! (With Eli Gould of Schering Corporation, 1967.)

aldehyde function at C-18. Because steroids with a substituent at
C-18 are not generally available and because the C-18 position is
unactivated and could not be activated by processes then
known, the synthetic problem was indeed a good one.

86   R = OH

87   R = H

After some thought, I proposed a solution, which was the good idea. From my earlier days of chlorinated hydrocarbon pyrolysis, I knew about the extensive work of E. W. R. Steacie, who later became director of the National Research Council of Canada, on the pyrolysis of organic nitrites. The postulated initial step was the cleavage of the O–NO bond into NO and an alkoxide radical. The required pyrolysis temperature would not be compatible with steroids, but I knew also of a limited amount of work on the photolysis of nitrites, about which there was some dispute concerning the formation of alkoxide radicals and NO.

One could hypothesize the reactions shown in scheme 10. By photolysis, the nitrite **88** gives an alkoxide radical (**89**) and NO; rearrangement of **89** affords the carbon radical **90**, which is captured by NO to give the nitroso derivative **91**. The isomerization to oxime **92** is a well-known process, and oxidative hydrolysis by nitrous acid would afford the aldehyde **93**, which would surely cyclize like a furanose sugar to give **94**. There was little precedent at the time for the rearrangement of **89** to **90**. But again, I knew from my earlier work on chemical kinetics that carbon radicals are very efficiently captured by NO.

The conformation of the steroid nucleus is perfect for the proposed reaction. The hydroxyl at C-11, which is postulated to become an alkoxide radical, is at the axial position, as is the C-18 methyl group. This 1,3-diaxial relationship is a guarantee that the groups are pushed together as required by the rearrangement of **89** to **90**.

Now we had the good problem and the good idea. The good experimentalist was John M. Beaton. He was a Scotsman from Glasgow who had done his Ph.D. with F. S. Spring on triterpenoid chemistry. Postdoctoral experience with Woodward produced the needed maturity. Beaton had been hired by the Schering Corporation to work in Bloomfield, but he was willing to return to Cambridge. In a few short weeks and with some technical assistance, he converted the readily available corticosterone acetate **95** into its nitrite **96** and photolyzed the nitrite by using 350-mμ irradiation to give the oxime **97** (scheme 11). Treatment of the oxime with nitrous acid gave aldosterone acetate **87**.[84]

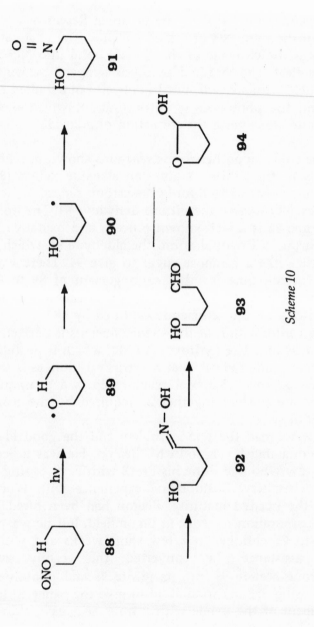

Scheme 10

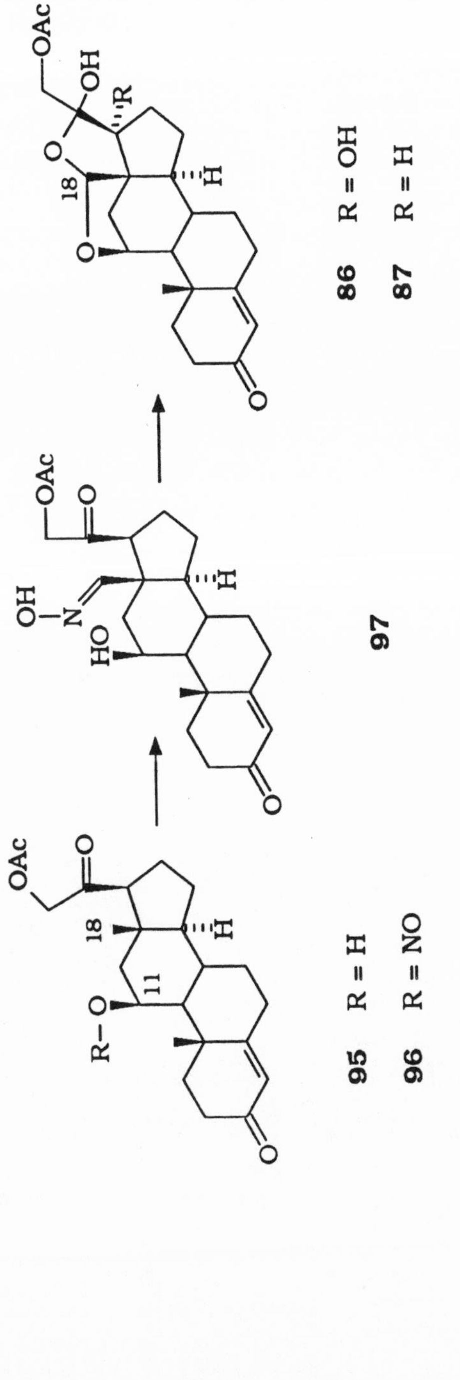

95 R = H
96 R = NO

97

86 R = OH
87 R = H

Scheme 11

Aldosterone acetate was obtained crystalline and completely pure. The yield was sufficient for the Schering Corporation to make 60 g without difficulty. I remember well a lecture that I gave at the 1960 spring meeting of the American Chemical Society, when I produced from under the desk a bottle containing 60 g of well-crystallized aldosterone acetate at a time when the supply elsewhere in the world was in milligram quantities.

I believe that our exploit with aldosterone acetate played a role in the foundation of the Woodward Institute in Basel, Switzerland. When we had our 60 g of the compound, the chemists at Ciba (as it was known then) produced an elegant, but very long, total synthesis of aldosterone. From a practical point of view, the RIMAC method was much superior. A management change at Ciba followed, and Woodward was invited to direct the new Woodward Institute. It was the same laboratory but with a new name. Excellent work was done for many years both at RIMAC and at the Woodward Institute. During these years, Woodward and I regularly crossed the Atlantic in opposite directions. Perhaps it would have been easier if I had gone to Basel and Woodward had stayed in Cambridge!

After the aldosterone acetate synthesis, Beaton left for the Upjohn Corporation, where he has had a distinguished career. At RIMAC, many aldosterone derivatives were made and improved syntheses were devised[85] that were all based on nitrite

At the IUPAC meeting in Prague, September 1962. Left to right, Carl Djerassi, R. B. Woodward, Barton, Vladimir Prelog, and V. Torgov.

photolysis as the key step for the substitution at C-18. Later work at RIMAC involved an improved synthesis of hydrocortisone, the development of electrophilic fluorination, and the synthesis of vitamin D metabolites. The work was greatly aided by R. H. Hesse, who became codirector of chemical research and participated fully in the conception of what was to be done.

Working in France

In 1977, at the age of 59, I was appointed director of research in the Centre National de la Recherche Scientifique (CNRS) at the Institut de Chimie des Substances Naturelles (ICSN) at Gif-sur-Yvette in France. A year later, I retired from Imperial College at the earlier retirement age and became the director of the ICSN. My action may be regarded as brave or foolhardy. In any case, I will thank Edgar Lederer and Pierre Potier for their friendly welcome.

For some years before the move to Gif-sur-Yvette, I had been a member of the directory committee of the ICSN. This group met once a year, listened to some scientific lectures, and gave an opinion, in principle, on all work done. Advice was sought, and encouragement was given. The food and drink were very good, and I always enjoyed it. When I knew that Lederer was retiring at the age of 70 in two years' time, I felt that I could be a candidate. Because there were then no living Nobel Prize winners in chemistry in France, I could be at least an ornament. I must explain that my wife is French and that I understand the

Hairy, 1975.

Specimen cupboard at Gif-sur-Yvette, 1985.

Flanked by T. S. Osdene and Frank Resnik at the Philip Morris Science Symposium in Richmond, Virginia, 1975.

French language perfectly and speak it well enough, with an English accent that is supposed to be charming. My wife was very pleased at the possibility of returning home after 20 years of exile in the United Kingdom, and the CNRS was very pleased to appoint me.

I was in for many surprises in France. The CNRS is a highly political organization and responds to each change in government. The directors of the CNRS usually are changed when the government changes, and all the presidents of the commissions are changed also. The commissions of the CNRS examine and judge the work of all the scientific workers employed by the CNRS and of the university groups that are funded by the CNRS. To someone from the Anglo-Saxon world, the CNRS seems a strange and unwieldy organization. However, when you know the history of French science in this century, you realize that the CNRS has made and still makes an essential contribution to French scientific life. The value of specialized institutes like the ICSN can be debated. The same work could be done more cheaply in the university system if one had a fair system for the distribution of the money. However, in France this fairness was not possible in the past.

The CNRS has rendered a great service to French science by making professors unhappy. Unhappy professors are creative professors, but they must not be too unhappy, or they may give up hope. A French professor cannot finance his research directly from university sources. He must seek funding for his group either from the CNRS (which is difficult) or from industry (which is very difficult). The commissions of the CNRS judge the applicant professors, and about one professor from every three or four who apply eventually obtains funding. Professors work very hard for years to get into the CNRS system, and once they are in, they have to work hard to stay in the system, because the commissions can also remove their support if they so judge. With this stimulating environment, it is not surprising that professors have a heartfelt ambition to abolish the CNRS and advocate the uniform redistribution of funds to the universities. Let us hope that the new French government will not pay heed to this siren song. France still needs badly the CNRS.

Retirement

When I went to France, I thought that I might retire at 70, like Lederer. However, he had served in the French army, and even more important, he had *une famille nombreuse*. I lacked in both respects, and so my retirement age was to be 68. However, the new Socialist government, who wanted to promote their supporters to be judges in the law courts, decided on a uniform retiring age of 65. Because I was already over 65, I ended up retiring at 67. It was much to the credit of the CNRS that they allowed me to continue at Gif-sur-Yvette for 15 months after my formal retirement. A great deal of work was accomplished in this short time.

I knew that I would enjoy living in France, but I was not certain if I would quietly prepare for real retirement or not. As soon as I realized that everyone expected me to retire quietly, I started working very hard again. In the nearly ten years that I spent in Gif-sur-Yvette, I accomplished as much as I did in the decade from 1950 to 1960, which was my previous high point. One of the stimulants was that money and good students were not automatically available, as they had been at Imperial College.

Not much would have been accomplished in France without the help of friends in industry. I am particularly grateful to my friend Jean Mathieu (of Roussel–Uclaf) who arranged significant funding for work in Gif-sur-Yvette. Without this help, I probably would not have accepted the position. An amusing aspect of my relation with Roussel–Uclaf was that I was asked at first to be a consultant. I accepted on the condition that they gave my fees to the ICSN for research. Later, I was surprised to be asked to become a member of the board of directors, and I was happy to accept because that position meant six meetings a year with six splendid lunches. However, this arrangement was too good to last, and when the Socialist government came to power, all the directors were removed and replaced with real Frenchmen. Because the company was doing well and went on to become even more profitable, I felt that this purely political action was unjust. Fortunately, I could not question the excellence of the man chosen to replace me, Pierre Potier, my codirector at ICSN. He has a much better understanding of the pharmaceutical industry than I have.

When I was nearing retirement in Gif-sur-Yvette, two American universities suggested that I might join them as a distinguished professor. Because I had been asked to leave the ICSN and no real British or French offer was forthcoming, I was very glad to say yes to Texas A&M University. College Station, Texas, is about the same size as Gif-sur-Yvette, and the major activity is directly or indirectly related to the intellectual life of the university. Of course, Paris is not 45 minutes away by subway. It is about 12 hours away by plane!

I did receive one other offer, but I am not certain how serious it was. My friend, the late Professor Ovchinnikov, director of the Institute for Natural Products Chemistry in Moscow, learned that I was leaving for Texas and suggested that I come to his Institute. He said that he would double whatever the Americans offered me. With an Institute of 1000 workers, recently doubled from 500 and soon to be 2000, he certainly had the means to carry out his suggestion. I thanked him very much but declined on the grounds that I do not speak Russian and that the winters are too cold.

Life After Retirement

It is too early to give any report on my scientific life at College Station, but I have the determination and the vigor to do something significant, and in addition, I am always lucky in chemistry. Most of my contemporaries are either dead or retired, but I am inspired by H. C. Brown who, at the age of 75, has just made an important and fundamental contribution to asymmetric synthesis with organoboron compounds. I can run faster than Brown can, but how will it be when I am 75? In the meantime, I am very happy to be a (new) Texan.

Organic chemistry is an experimental subject, and except for the occasional theoretical concept, like that of conformational analysis, it is done by teamwork. I have been fortunate in my life to have had many remarkable collaborators whose work and suggestions I have greatly appreciated. Covering the period and

Teaching the fifth grade (Texas, 1988).

subjects of this article, I cite some of my colleagues in alphabetical order, indicating the main field of study in which we collaborated: M. Akhtar, University of Southampton (nitrite photolysis); J. E. Baldwin, University of Oxford (nonadrides); J. M. Beaton, Upjohn Corporation (aldosterone synthesis); D. S. Bhakuni, Lucknow Government Laboratory (alkaloid biosynthesis); C. J. W. Brooks, University of Glasgow (steroids and triterpenoids); A. W. Burgstahler, University of Kansas (caryophyllene); H. T. Cheung, University of Sydney (clerodin); T. Cohen, University of Pittsburgh (biosynthesis of alkaloids); A. D. Cross, Syntex and Zoecon (clerodin); O. E. Edwards, National Research Council of Canada (usnic acid); (the late) D. Elad, Weizmann Institute (columbin); A. K. Ganguly, Schering–Plough Corporation (photochemistry); J. B. Hendrickson, Brandeis University (fuscin); R. H. Hesse, RIMAC (alkaloid biosynthesis); R. B. Kelly, University of New Brunswick (lanosterol); A. S. Kende, University of Rochester (photochemistry); J. Klein, Hebrew University of Jerusalem (photochemistry); G. W. Kirby, University of Glasgow (alkaloid biosynthesis); J. E. D. Levisalles, University of Paris (photochemistry); P. D. Magnus, University of Texas, Austin (tetracycline); F. McCapra, University of Sussex (long-range effects); Alex Nickon, Johns Hopkins University (caryophyllene); K. H. Overton, University of Glasgow (triterpenoids); J. T. Pinhey, University of Sidney (photochemistry); S. K. Pradhan, Bombay (limonin); P. G. Sammes, Brunel University (photochemistry); W. Steglich, University of Bonn (salutaridine); S. Sternhell, University of Sydney (limonin); Prof. J. K. Sutherland, University of Manchester (nonadrides); J. B. Taylor, Rhône–Poulenc (alkaloid biosynthesis); W. C. Taylor, University of Sydney (photochemistry); and B. J. Willis, Quest International (hindered olefins).

I have a special acknowledgment to make to A. I. Scott of Texas A&M University. During the Birkbeck and Glasgow days, we were close collaborators on aldosterone and on the intricate chemistry of geodin and erdin. This collaboration led to his lifetime interest in biomimetic chemistry, which started with his brilliant synthesis of griseofulvin and has continued to the present with outstanding work on vitamin B_{12} biosynthesis and high-field-NMR studies of enzyme mechanisms. We are together again, not in the same relationship, but in one of equality in a common life interest in organic chemistry.

Herbert C. Brown and his son, Charlie (without his water pistol!). (Photograph courtesy C&E News)

I have not cited all the names that I could have; I mentioned only some professorial colleagues or those who have been especially successful in industry or government. I believe that the persons I have cited have something in common—a devotion to excellence in organic chemistry and a marked ability to pick the significant problems of the day to work on. Perhaps contact with me helped some of them.

Coda

Twenty More Years of Discovery, 1970–1990

The respected editor of this volume has persuaded me to write a coda. Otherwise, he suggests, the reader will think that I have done nothing in the last 20 years. The late Sir Christopher Ingold was very fond of such closing statements, but it took me many years before I slipped one into the literature.[86]

Since about 1970, I have concentrated largely on the invention and discovery of new chemical reactions. In the invention of a chemical reaction, one must first select a reaction that is needed. Most academic professors do not interest themselves in this type of work and prefer to discuss, with great erudition, reactions that are already known. Our friends in industry are most aware of the need for new reactions that give higher yields under milder and less costly conditions than before.

However, the accidental discovery of a new reaction, a much more common event, is still welcome. I start this section with an amusing accidental discovery.

The Game-of-Bridge Reaction

In 1970, Selim Achmatowicz, the son of one of Poland's most distinguished professors of organic chemistry, came to work with me for a year at Imperial College. He brought with him his own stipend. I asked him to study methods for the synthesis of the thiobenzoate (98) of cholesterol (99). I was planning a study of the interchange of functional groups on heating that we later named the α,ω-rearrangement.[87] Its origins lay in the $5\alpha,6\beta$- → $5\beta,6\alpha$-dibromocholestanol ester equilibration that we had studied in conformational terms many years before.[88] An extension had already been made,[89] and the final generalization was then foreseen.[87]

A beautifully yellow solution of the thiobenzoate (98) was eventually prepared in ether. Achmatowicz was a very good bridge player, so good that he was a member of the Polish team and often played in international tournaments. His yellow solution was prepared on a Thursday afternoon, just before he left on Friday to play in a weekend tournament. There was no time to work up the solution and isolate the yellow thiobenzoate (98).

It was my habit to visit the laboratory at least once a day. On Friday and Saturday, I noticed that the bright yellow color of the solution of (98) was fading. By Monday morning, the yellow color was almost gone, and when Achmatowicz returned on Tuesday morning, the solution was colorless.

We had discovered a new and almost quantitative photochemical reaction (reaction 7). The products of the reaction were cholestal-3,5-diene (100) and thiobenzoic acid.[90] In collaboration with Sir George Porter and J. Wirz, we studied the mechanism of

$$R-O \qquad \longrightarrow \qquad \tag{7}$$

98   R = C_6H_5CS

100

99   R = H

With Linus Pauling at the Stadttheater, Nobelpreistrager—Tagung, 1977.

this reaction (elimination from the lowest $n \rightarrow \pi^*$ triplet).[91] The reaction has a high quantum yield. The limiting factors in this elimination were studied, and it was shown that the hydrogen abstracted by the thiocarbonyl radical had to be allylic or benzylic or in a similar relationship.[92] Without the game of bridge, the ethereal solution would have been worked up in the usual way, and an excellent new reaction would still be waiting to be discovered.

Radical Deoxygenation—The Reaction of Barton and McCombie

The reaction of Barton and McCombie[93] was invented. From my contacts with the Schering–Plough Corporation, who had important interests in the chemistry of the aminoglycoside antibiotics, I knew that the removal of certain secondary hydroxyl groups from antibiotics actually increased the biological activity. Biological activity rose because the bacteria could no longer detoxify the antibiotic by acylation or phosphorylation of the

With George Porter, later President of the Royal Society (1970).

secondary hydroxyl groups that had been removed. Because ionic methods of removing hindered hydroxyl groups have certain limitations that are irreparable when neighboring-group participation is concerned, a radical deoxygenation process would be much superior and also original. However, the C–O bond in a secondary alcohol is a strong bond.

I at once thought about the game-of-bridge reaction. The overall driving force for this reaction was the change from the high-energy thiocarbonyl group to the low-energy carbonyl group. One could conceive that a thiocarbonyl derivative **101** of a secondary alcohol **102** could react with a tributyltin radical to give an intermediate radical **103**, which at a certain temperature would fragment to give the desired radical **104** and a fragment

105 containing the carbonyl group. In the presence of tributyltin hydride, the secondary radical **104** would certainly be quenched to give the desired deoxy compound and a tributyltin radical (scheme 12.) Thus we would be back to chain reactions again. The group X could clearly be $-C_6H_5$, $-S-CH_3$, imidazole, etc.

At this point, the good problem and the good idea again met the good experimentalist in the form of S. W. McCombie—Stu to his friends. Stu is not only a good experimentalist; he has an excellent knowledge of organic chemistry and a flair for knowing how to run a reaction. The overall combination was excellent, and within a few weeks a high-yield deoxygenation was achieved. We applied the reaction to a number of carbohydrate problems. For the first time in my life, I became a carbohydrate chemist.

The game-of-bridge reaction also gives excellent yields, but as far as I know, no other group has ever used it. On the

Receiving an honorary knighthood for wine tasting in Bourgogne, France, 1981.

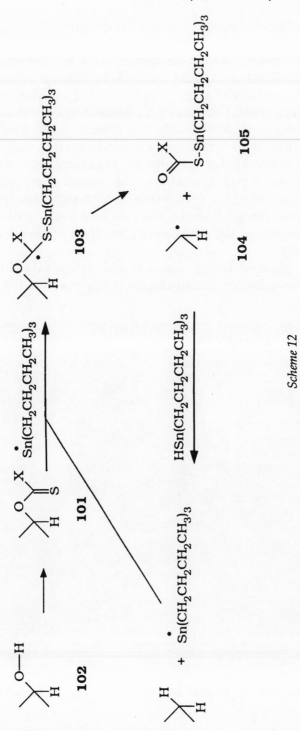

Scheme 12

contrary, the deoxygenation reaction was taken up with enthusiasm, and there are by now many hundred examples[94] of its application in the original or in modified form.[95] The difference is that the deoxygenation reaction was needed; the game-of-bridge reaction was not.

In the design of chemical reactions, the recognition of driving forces constitutes an important part of the thinking process. In the deoxygenation reaction, one driving force is the shift from the thiocarbonyl group to the carbonyl group. Other driving forces are the affinity of the tin radical for sulfur and the increase in entropy resulting from the fragmentation of 103 to 104 and 105 (scheme 12). Unless the temperature of the reduction is high enough, radical 103 can be reduced in competition with radical 104.

The Disciplined Radical

After a brief study[96] of the radical deamination of aminoglycoside antibiotics, which became a very efficient process, I turned to the carboxyl function. The carboxyl function is present in many biologically important natural products, such as biotin, all the prostaglandins and leukotrienes, and peptides and derivatives. It seemed to me that it should be possible to invent a method for converting $R-CO_2H$ into the radical $R^{\cdot}$. Since the days of Kolbe in the last century, the decarboxylation process $R-CO_2H \rightarrow R-CO_2^{\cdot} \rightarrow R^{\cdot} + CO_2$ has been known. However, the radical $R^{\cdot}$ produced in this way never gave very high yields in synthetic processes. I realized that the mixed anhydrides of thiohydroxamic acids and carboxylic acids (106) had a relationship to the kind of thiocarbonyl derivatives that we had used before in the radical deoxygenation of secondary alcohols. The comparison of 107 and 108 makes this relationship very clear. It is, in fact, better to talk about thiohydroxamic esters by treating the N—OH function as the equivalent of an alcohol. Their chemical properties in the ionic sense correspond to the ester group. They are slightly electrophilic, neutral compounds.

With the invention of a reaction in which a strong C—O bond in 107 is broken (reaction 8), it is self-evident that the weak N—O bond in 108 will break even more easily (reaction 9).

Thinking again in terms of tin hydride chemistry, I predicted the reaction **106** → **109** → **110** → **111** → **112** (scheme 13). It worked beautifully[97] and gave high yields of the decarboxylated "nor" hydrocarbon **112**.

The first student who worked on this project, David Crich, did not appreciate (nor did his laboratory colleagues) the sulfurous odor emanating from the normal preparations of thiohydroxamic acids. Crich showed unusual initiative and diligence in searching the chemical catalogs and found that the sodium salt of N-hydroxy-2-thiopyridone (**113**) was available (under a misleading name) from Fluka. To make the situation even more curious, the compound was not mentioned at all in the index of the catalog. In reality, this sodium salt is available inexpensively from the Olin Corporation as a 40% aqueous solution. One only has to acidify the aqueous solution with concentrated hydrochloric acid, and the highly crystalline thiohydroxamic acid **113** precipitates. But at the beginning, we did not know that.

From thiohydroxamic acid **113** (or its sodium salt), it was easy to prepare a wide range of the corresponding esters (**114**) for application of the tin hydride reduction procedure.[97] In the course of this work, a new decarboxylative rearrangement was discovered. Thus, heating compounds of type **114** without tin hydride leads to a smooth formation of **115** with the loss of CO_2. This reaction is another radical chain reaction in which **114** affords thermally the radical R'·.

This observation was significant, because it suggested that the radical R'· could be captured by other radical traps. Indeed, the addition of a thiol gave R'–H in excellent yield. The esters **114** are sensitive to ordinary tungsten light. In the absence of an added trap, they furnish the same derivatives (**115**) as those produced by the thermal process. This result means that reactions can be carried out over a wide temperature range from −60 °C upward.

Radicals generated from thiohydroxamic esters can be used in a large variety of chemical reactions and usually give high yields of products.[98] Why do these reactions work so well in synthesis? In my opinion, the radicals have been disciplined by a disciplinary group. In the deoxygenation reaction, the radicals are disciplined by the weak Sn–H bond. In the thiohydrox-

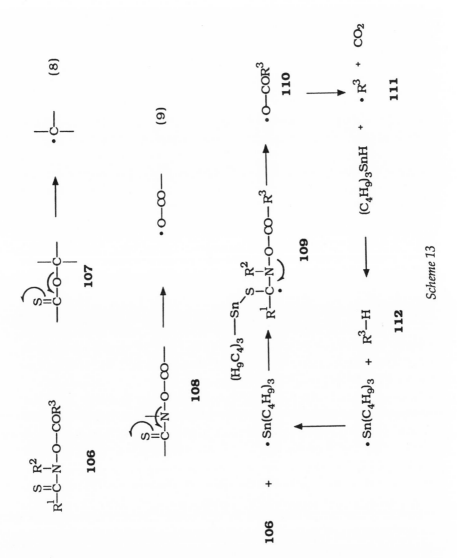

Scheme 13

113   R = H

114   R = R'CO

amic acid esters, the thiocarbonyl group is the ultimate disciplinary group. Radicals in these circumstances do not attack the solvent, and so many solvents like tetrahydrofuran, toluene, and dimethyl formamide can be used without reducing yields.

We have designed reactions such that the rate constants fit together. However, this statement does not help you much to think of new reactions; the concepts of the disciplinary group and the disciplined radical do.

The photolysis of thiohydroxamic acid esters of type **114** can be carried out in a flow reactor so that carbon radicals are detected directly without the need for trapping.[99] There is a spectacular reproduction of the ESR spectrum of a benzyl radical.

I previously predicted that irradiation of a thiocarbonyl function would afford radical chemistry.[100] The acyl xanthate group **116** cleaves smoothly into acyl radicals and xanthate radicals on irradiation with a tungsten lamp. The acyl radicals eventually lose CO if the R group is benzylic or *tert*-alkyl, and the final product is an alkyl xanthate **117**. Zard[101] recently showed by an ingenious series of experiments that the reaction is really a radical chain reaction with the propagation step as indicated for structure **118**. The radical R• can then be used in synthetic chemistry just as it can when it is derived from a thiohydroxamic ester. This new development is promising.

116   **117**   **118**

The Third Mechanism

The selective substitution of saturated hydrocarbons is a splendid challenge for the remaining years of this century. We started our own work on this problem in 1980.[102] J. I. G. Cadogan, who recently had been appointed director of research at British Petroleum (BP), a most prestigious position, came to see me at Gif-sur-Yvette to tell me that BP was planning to give research funds for a limited number of blue-sky projects from imaginative chemists of any age. Would I like to propose a project? Because I am always young (in spirit), I proposed the application of iron chemistry to the selective substitution of saturated hydrocarbons. The project was accepted, even though it was a rash proposal, and it turned into an extraordinary and bizarre chapter of original chemistry.

I thought that I would, at least at first, imitate the P450 enzyme mechanism, even though this mechanism was known to involve radicals. An article by Tabushi[103] on the oxidation of adamantane (119) had just appeared. Using an iron complex as catalyst, β-mercaptoethanol as reductant and ligand, oxygen as oxidant, and pyridine as solvent, Tabushi obtained less than 1% of oxidation, but the selectivity C^2/C^3 was 2.3, not like that for an oxenoid radical reaction ($C^2/C^3 = 0.1$). C^2 is the amount of secondary-substituted products produced, and C^3 is the amount of tertiary alcohol produced.

We repeated the work of Tabushi, as well as many comparable experiments, always with miserable results. One day, I was reading a general article about the primitive world with its reducing atmosphere. It occurred to me that most of the iron in the primitive world must have been metallic iron. The major iron ore deposits are, after all, supposed to have been produced by biological oxidative processes. I asked Michel Gästiger, who was doing the work, to add some iron powder (and acetic acid) to the Tabushi system. The result was a dramatic improvement in the percentage of oxidation (an increase of five- or tenfold!). We arrived at two systems for the selective oxidation of saturated hydrocarbons: Gif^{III}, Fe^0–pyridine–CH_3CO_2H–O_2, and Gif^{IV}, Fe^{II}–pyridine– CH_3CO_2H–Zn^0–O_2. With the Gif^{IV} system, an oxidation of 20% and a C^2/C^3 selectivity of 10–20 can be secured.

When we reduced the oxygen pressure to about 5%, we were surprised to find that the selectivity increased markedly while the total percentage of oxidation remained constant. Further investigation showed that the tertiary position in adamantane gave radicals that were competitively quenched by partition between oxygen and pyridine. No such partitioning occurred for radicals at the secondary position, and no secondary pyridine-coupled product was formed. The overall C^2/C^3 selectivity, with the pyridine-coupled products taken into account, was 1.2.

If secondary and tertiary radicals were both involved in this oxidation process, the amount of oxidized product should diminish with diminishing oxygen pressure, but the selectivity should not change. I concluded, therefore, that the mechanism of oxidation at the secondary and tertiary positions was different.[104]

It would be very strange if two different reagents were present, with one reagent attacking the tertiary position in a radical mechanism and the other attacking the secondary position in a nonradical way. A much more reasonable hypothesis is the presence of one reagent that attacks both secondary and tertiary positions, with the bond formed at the tertiary position then evolving into radicals.

My first hypothesis[104] was that an iron–carbon σ bond was formed at both positions, but the bond in the tertiary position evolved into radicals; a precedent[105] was available for this transformation.

To test the hypothesis, the secondary and tertiary radicals were prepared separately by thiohydroxamic ester chemistry,[106] as described previously. Both radicals partitioned between oxygen and pyridine in the same way. In addition, the secondary pyridine-coupled products could be separated by gas–liquid chromatography (GLC) from their tertiary analogs. So the original GLC–mass spectrometric analysis was confirmed.

Finally, the oxidation of saturated hydrocarbons can be carried out efficiently by using the cathode of an electrochemical cell.[107] The oxidant for iron is superoxide, and the iron is surely Fe^{II}.

We have carried out many experiments to trap the postulated iron–carbon and iron–carbene intermediates. Recently, we found a reagent, diphenyl diselenide, that, under Gif^{III} conditions, very efficiently traps an intermediate that we consider to be the iron–carbon σ bond.[108] In this reaction, no oxidation of the hydrocarbon is seen, but the tertiary (120) and secondary (121) phenylselenoadamantanes are formed in more than 30% yield with a quantitative recovery of the hydrocarbon. In a similar reaction with cyclohexane, phenylcyclohexyl selenide is formed. The use of a large excess of cyclohexane relative to diphenyl diselenide permits more than 70% of the phenylseleno residues to be incorporated into phenylcyclohexyl selenide, and the rest of the selenium is converted into 2-phenylseleno-pyridine.

(tertiary)

R^1 (secondary)

R^2

119    $R^1 = R^2 = H$

120    $R^1 = C_6H_5Se$; $R^2 = H$

121    $R^1 = H$; $R^2 = C_6H_5Se$

119

Saturated hydrocarbons are totally inert toward diphenyl diselenide, and there is no reaction without all the components of the Gif[III] system. Both iron and oxygen are essential for the reaction. The Gif[IV] system permits a competition between the ordinary oxidation reaction and the phenylselenation reaction. In all the work with phenylselenation, the overall C^2/C^3 selectivity is about 1.2, reflecting the same initial process (iron–carbon σ bond formation).

There are two well-known mechanisms for the selective oxidation of saturated hydrocarbons. The Fenton reaction of H_2O_2 with Fe^{II} leads to the production of hydroxyl radicals.[109] The second mechanism is the iron-based porphyrin system of the P450 enzymes. On the basis of this system, alkoxy-radical-like chemistry is seen in porphyrin models.[110] Saturated hydrocarbons are attacked with a selectivity normal in radical reactions

Presenting the Tetrahedron Medal to Robert Maxwell, the publisher, in Oxford, 1983. A very nice way to wish him a happy 60th birthday!

(tertiary > secondary > primary), sulfides are oxidized to sulfoxides, and olefins are epoxidized.[111,112]

The Gif family of systems oxidizes saturated hydrocarbons to mainly ketones, and the order of selectivity is secondary > tertiary ≥ primary. The Gif systems do not oxidize sulfides to sulfoxides (or reduce sulfoxides to sulfides), nor do they epoxidize olefins. Olefins are oxidized to unsaturated ketones, and the rate of oxidation is similar to the rate of oxidation of the corresponding saturated hydrocarbon.

The Gif systems clearly constitute the third mechanism of selective oxidation of saturated hydrocarbons. On the basis of less-decisive work on penicillin cyclase[113] and soyabean lipoxygenase,[114] it seems that Nature may be using this mechanism also. The story is not finished; perhaps it has hardly begun. Aerobic life depends on the iron–oxygen–substrate system and could not exist otherwise. Perhaps there are other mechanisms. I hope that I can live long enough to participate in their study.

Coda to a Coda
My Philosophy

By special request from the editor and the referees, I have been persuaded to expand upon my coda. It seems that my overall philosophy about life needs to be expounded.

My early childhood years were happy but disciplined. Meals were served to the minute, and late arrival was punished. The first horrible experience of my life was to be sent off to a boarding school at the age of 11. Young boys are always cruel to each other, and only physical strength can protect you. It was a new world for me.

The headmaster was a priest, and he took pains to shape the more responsive boys into his image. By the age of 13, I was ready for Greek and Hebrew and the priesthood, but my father decided that I must go to a better and larger school, Tonbridge. The atmosphere at Tonbridge was different, and the food was much better. My frugal life was over. Moreover, as manhood approaches, boys are less cruel to each other.

The worst feature of life at Tonbridge was the rule that dormitory windows were left open day and night all year round. There was no central heating, and we shivered in bed.

Towards February each year, I contracted bronchitis and influenza and had to be sent home. I did not return until the summer term. The long convalescence was spent at the seaside. Without any parental influence, I started to spend my pocket money on heavy and learned books on science and philosophy. My enjoyment of such books indicated that I had already chosen the academic path.

The vocation of the priesthood lingered with me through the adolescent years. Certainly, I felt that I must give my life to some noble cause. Eventually this noble cause became science in general and chemistry in particular. The discipline of constant study came easily to me, and from the age of 18 onwards, I have not stopped. This discipline must have been the result of the early influence of home life and of the headmaster–priest, who

Happily, there was no sequel (1924).

had splendid sermons on the bad things that happened to boys who did not work.

As a student at Imperial College, I naturally took to precision in the writing of formulae and to an appreciation of logic in a scientific argument. As I have already mentioned, when I began doing research, the goals were defined by the wartime circumstances, but in my spare time I started some library work. I believe that I taught myself how to think beginning at that time. Trying to see order in a chaotic mass of conflicting facts suits me well.

When I began independent research in 1945, I picked problems that I thought were worth solving, but not problems that were too difficult to be solved within a given time scale. The posing and solving of a problem gives personal satisfaction, which is increased if the solution is elegant and if the problem could not be solved by others. When I saw the relationship between the pyrolysis of chlorinated hydrocarbons and *cis* elimination in steroids, I was satisfied. The same satisfaction came when I saw the relationship between conformation and reactivity in steroids. I particularly enjoyed deducing that the accepted formula for Pummerer's ketone was wrong, because I then foresaw the true biosynthesis of morphine. As an aside, it is amusing that Sir Robert Robinson at the Natural Products Meeting of the International Union of Pure and Applied Chemistry in Prague in 1962 gave a talk in which he still used the wrong formulae. At the end of the talk I could not help pointing out to him (personally) that his formulae were some years out of date. He was very angry, but he did not hit me with anything. In any event, he quickly forgave me.

As soon as I gave up structural work as a major occupation, I found my true role in chemistry as an inventor of chemical reactions. Although, like an artist, I seek elegance and personal satisfaction, I am still pleased when I do something useful. I realize that there is a direct relationship between the utility of chemistry and how much academic research can be funded. It is strange that the same restrictions do not seem to apply for physics or molecular biology.

I have looked all my life for the significant problems in natural product chemistry that required an imaginative, intellec-

Two Bartons: Sir Derek with fellow chemist Jacqueline K. Barton of Caltech, Bürgenstock, 1989. (Photograph courtesy K. Zimmermann)

tual solution. New reactions and, to a less extent, new reagents have given me the satisfaction that I need. The synthesis of aldosterone by the then new reaction of nitrite photolysis was such a case. Before the next significant reaction could be invented, there had to be a change in my personal life. Eventually I recovered, and when the problem of the deoxygenation of aminoglycoside antibiotics was posed, I was ready with the solution. Again radical chemistry was involved.

My philosophy of chemistry has always been to read very widely and to attend lectures on other branches of chemistry. The problems of your neighbor may be simple to you, if you have prepared your mind correctly. Even so, it is amazing how long it may take to see the relationship between related reactions. Our new radical-generating system from carboxylic acids is closely related to the deoxygenation reaction. I could have deduced its existence in 1975 and not waited until 1981.

I never forget, either, the accidents that lead to new reactions. Accidental discoveries are not really as satisfying as solving a problem, but at least, one can look back on how one realized what these accidents implied. Many people do not read enough, so that often they cannot distinguish between what is original and what is already known (and trivial).

I have always tried to avoid following the flock. I have worked in many fields, but as soon as these fields became popular, I have moved on. I have made the joke of saying that if you

With David Ollis at the EUCHEM Conference on Stereochemistry at the 1989 Bürgenstock Conference. Rolf Huisgen is on the far right. (Photograph courtesy K. Zimmermann)

cannot remember all the published papers in the field you are working in, then it is time to move on.

Asking the right questions, choosing the significant problems, and critical thinking are all part of the creative process. Do you need more? Yes, in the world of experimental chemistry, you do need more, especially if you have administrative responsibilities. Interaction in the real world requires a certain sense of politics. I must have this sense because I was responsible for administrative units from 1950 until 1985—35 years! I am glad to write that I never had a serious disagreement with anyone during this period. The secret was to set a good example of application and seriousness of purpose while at the same time being flexible. An administrator must avoid being maneuvered into a corner. Now it is nice not to be responsible for anything else but good science. May it last for many years.

With E. L. Winnacker at the EUCHEM Conference on Stereochemistry, Bürgenstock, 1989. (Photograph courtesy K. Zimmermann)

Coda to a Coda to a Coda
Reminiscences of Friends and Colleagues

Our respected editor has asked me to add something more about the various personalities in the world of chemistry whom I have known. I do so with some reluctance because when writing of the living— however much you praise them— it is never enough. So, you can only add to your list of enemies!

Sir Robert Robinson

By the time I knew him well, he was already near blindness. He was lucid to the end. I remember a meeting of the *Tetrahedron* editorial board where we discussed the final format for *Comprehensive Organic Chemistry* (COC). My views finally prevailed over his. (He preferred a kind of Dictionary approach.) As compensation I suggested that Sir Robert write an introduction to *COC*. He agreed at once.

A week later he rang me up to tell me how much he appreciated the invitation to write the introduction. In a clear

Sir Robert and Lady Robinson, October 1953. Au revoir, not yet adieu. (Photograph courtesy C&E News)

voice he asked if he could write a long introduction, say 30 pages. I replied, "Of course, Sir Robert, anything you wish." He said, "Good, I want to write all about the electronic theory of organic chemistry and of the secondary role played by [Sir Christopher] Ingold."[115] Two days later, at the age of 88 years, Robinson was dead.

Sir Christopher Ingold

Sir Robert and Sir Christopher were very different. When Sir Robert became famous, he could never say no to an invitation. He was too busy to think carefully or to read the literature. As a result, his lectures were confused and confusing because they were not properly prepared. In contrast, Sir Christopher's lectures were impeccable. It was as if he were preaching the gospel. The introduction was followed by the logical presentation of theory and results blended into a harmonious whole, neatly summed up in the conclusion. One felt that one understood perfectly and that no better, or more qualified, theory could follow. References to past or current literature were nonexistent, except for that of Finkelstein, whose reaction was quoted repeatedly.

I admired Sir Robert for his past and Sir Christopher for what, to me, was the present. Sir Christopher did not believe in radical reactions. He thought they were messy. Therefore, I appreciated it very much when he encouraged me in my own radical chemistry. Sir Robert, however, was happy to approve of the efforts of Hey and Waters to demonstrate the reality of radical chemistry, even if messes resulted. How wise of Keith Ingold, Sir Christopher's famous son, to concentrate on radical chemistry!

Carl Djerassi

My first trip to the United States took place in September 1949 and I had arranged to spend one night in the Harvard Club in New York, before proceeding to Boston. I had corresponded with Carl about steroid chemistry before my arrival, and we agreed to meet that evening. He drove over from New Jersey in

With my wife, Christiane, Carl Djerassi, and his wife, Diane Middlebrook, at the IUPAC 11th International Symposium on the Chemistry of Natural Products, Golden Sands, Bulgaria, September 1978.

the afternoon to pick me up for dinner at Charlie Huebner's house. At that time both worked for Ciba, although the next day Carl started his drive to Mexico City to begin his famous work with the Syntex Corporation.

Carl was, and still is, an exceptionally intelligent and gifted person. In the car he talked to me at the same speed at which he drove—which was very fast. I did not have much chance to reply. After 30 minutes of this, I became depressed. I thought, how could I compete with Americans if all of them were like Carl Djerassi! Happily, or otherwise, very few Americans can keep up with Carl's extraordinary vitality. So, as the months went by, I recovered from my depression.

Sir John Cornforth

I have known Kappa for many years. He has a fine mind, and Sir Robert's synthesis of cholesterol would never have been finished without him. Cornforth's Nobel Prize was a just reward for the beautiful and elegant work that he did on the biosynthesis of cholesterol. He owes a lot, both chemically and personally, to his charming wife Rita. Kappa also writes clever limericks. I particularly appreciated the one that he wrote for my 60th birthday, where he managed to rhyme 'Barton' with 'French blonds by the carton'. Delicious, I agree.

Arthur J. Birch

Arthur Birch was at Oxford the same time as Kappa and Rita Cornforth, but his relationship with Sir Robert was less cordial. Arthur likes to tell the story about when Sir Robert showed him the original Dupont patent of Wooster on the metal–ammonia reduction of benzene to 1,4-dihydrobenzene. During World War II Sir Robert was a consultant to ICI, and there was an agreement between Dupont and ICI to exchange information and not to work on the same subjects, so this information was confidential. Arthur, of course, had, and has, a wonderful appreciation of significance in organic chemistry. I think that I share this capacity, because I immediately understood the importance of Arthur's first publications on what is called the Birch reduction. He realized at once the potential importance of the reaction in natural products chemistry in general and in the synthesis of 19-norsteroids in particular.

Sir Robert was away from Oxford during much of the war on work of national importance, but every six months or so, he would ask Arthur how his work was going. Arthur was supposed to be working on ideas proposed by Sir Robert. Once he had seen the 1,4- dihydrobenzene reduction, however, Arthur went to work on similar chemistry with enthusiasm. When he told Sir Robert what he had accomplished, Sir Robert was very angry and told him to stop at once. ICI also told him to stop at once, because it was in violation of the Dupont agreement. Naturally, Arthur did not stop at all—but he no longer told Sir Robert what he was doing.

After the war, Arthur moved to Cambridge. There, he proposed his polyacetate theory of the biosynthesis of many natural products, an excellent theory on which Arthur performed pioneering experiments after his return to Sydney, Australia. However, Sir Robert pointed out the much earlier theory of Collie who had proposed the same biosynthesis for a more limited group of aromatic natural products. Besides the Birch reduction and the polyacetate theory of biosynthesis, Arthur was also a pioneer in the application of organometallic chemistry to organic synthesis, especially natural product synthesis.

I was very pleased when the board of editors of Tetrahedron Publications awarded the prestigious Tetrahedron Prize to Arthur for his contributions to creativity in organic chemistry.

Gilbert Stork

When I arrived at Harvard in 1949, Gilbert was working with his own hands on the synthesis of morphine. This was an ill-advised activity. Gilbert had, and still has, a brilliant mind.

Following Japanese tradition, Tetsuo Nozoe, Barton, Gilbert Stork, and V. Herout sample sake from a barrel at the 16th ISCNP IUPAC Meeting in Kyoto, June 1988.

However, its extension to the control of his hands was somewhat lacking, so he did not accomplish very much until his first graduate students arrived. After that, his career, characterized by exceptionally imaginative chemistry, flowered. He was also awarded the Tetrahedron Prize as well as sharing a Roussel Prize with Ron Breslow.

Vlado Prelog

Vlado is an unusual man—a master of many tongues and of many anecdotes, it is possible to underestimate him. He has, in fact, a profound intelligence that, in combination with the sharp minds of Sir Christopher and R. S. Cahn, gave us a system of specifying chirality that has stood the test of time. His pioneering contributions to adamantane chemistry, alkaloid chemistry, biotechnology, antibiotics, and conformational analysis earned him the Nobel Prize that he shared with Kappa Cornforth. At 84, Vlado seems as lively as ever.

Albert Eschenmoser

Albert and I first met in print over the chemistry of caryophyllene. I recognized at once that a fellow intelligence had arrived. His contributions to the theory of a coherent biosynthesis of tetra- and pentacyclic triterpenoids, along with those of Duilio Arigoni, Oskar Jeger and the great Leopold Ruzicka, were a masterpiece of (correct) hypothesis.

The career of Albert has progressed in brilliant, logical steps. Realizing that structure determination was too simple, he concentrated on synthesis and then biosynthesis. From onocerin through the synthesis of Vitamin B_{12} (with R. B. Woodward), and the elegant photochemical synthesis of B_{12} (without Woodward), to the theory of the biosynthesis of corrins, and then of life itself, Albert has never made a mistake. The enormous and profound work on B_{12} synthesis could fill a book and is deserving of one. It was very fitting that Albert was awarded the first Tetrahedron Prize.

Albert Eschenmoser speaking at the EUCHEM Conference on Stereochemistry, Bürgenstock 1989. (Photo courtesy K. Zimmermann)

J. (Jack) E. Baldwin

Although I have made mistakes in chemistry, I have seldom misjudged those who have worked with me for any period of time. When Jack was an undergraduate at Imperial College, I recognized at once that he was someone unique. His capacity to wreck cars seemed to be matched only by his brilliance as a chemist. He worked on a difficult but interesting problem (the structure of the nonadrides) for his doctorate. At that time, I was able to appoint him to a junior position at Imperial College, hoping that he would climb the academic ladder as fast as possible. However, without publishing much, Jack had already achieved fame through the academic grapevine. Seduced by a fast car and a significant salary increase, he moved to Penn State University. His brilliant career took him next to MIT. Soon, he was on his way to King's College, London. There, he overplayed his hand by threatening to leave if his demands were not met—and the Principal accepted his resignation. Back again at MIT, it seemed that Jack might never return to the United Kingdom. However, my retirement from Imperial College precipitated negotiations, but Oxford intervened and was fortunate to attract him to that would-be sleepy institution. Jack has been very good for Oxford. He has collected money, instruments, and good students, and he has shaken up and inspired the staff members. In spite of Jack's various threats to leave because of the inadequate United Kingdom salaries, he is still there, enjoying the excellent scientific and human facilities of Oxford.

Jack's chemistry is never trivial and often brilliant. He has found his own Holy Grail in the biosynthesis of penicillin, a deceptively simple molecule assembled by nature in an ingenious and highly original way. He postulated a role for the iron–carbon bond, difficult to prove by enzymatic work, but more readily obvious in our studies of the iron–carbon bond in Gif-type chemistry. Jack is fortunate in having found an exceptional wife, who actually understands him.

Duilio Arigoni

I first met Duilio at a terpene meeting that I helped to organize at Glasgow University in 1957. He arrived with Oskar Jeger. It was the beginning of the final determination of the structure of limonin and also of the saga of santonin photochemistry. Our first paper on the latter subject was already in press.

Duilio Arigoni at the EUCHEM Conference on Stereochemistry, Bürgenstock, 1989. (Photo courtesy K. Zimmermann)

Duilio is brilliant at any time, on any subject, and in any language. His expositions are crystal clear and he makes complicated biosyntheses seem simple. His reputation is outstanding, but it is based more on his lectures than his publications, which are not numerous. He really needs an editor to organize his work for him. His most important research is probably the synthesis and use of the chiral methyl group in studies of biosynthesis. How lucky the Eidgenossische Technische Hochschule has been to have Duilio Arigoni, Albert Eschenmoser, Oskar Jeger and Vlado Prelog at the same time. Of course, all were appointed by the great Ruzicka, who had the same flair for scientific talent as I think I have.

Paul deMayo

Paul deMayo came to see me when I was working at Birkbeck. He was a refugee from University College Medical School, where he had been working as an assistant. He had taken his bachelor's degree at Exeter, which at that time was part of the external side of London University. The teaching at Exeter then was not inspiring, and Paul did not get a first-class degree. I was able to secure a fellowship for him from the Colonial Products Research Council, thanks to my cordial relations with the director, Sir John Simonsen. With this he began work on his doctorate.

I quickly realized that Paul was especially gifted, and together we carried out a number of structural determinations. Especially notable was our work on the toxic triterpenoid icterogenin. When we realized that cross-conjugated dienones were very active photochemically in a series of unprecedented rearrangements, we began to study the photochemistry of santonin. A few years later, Paul became a Professor at the University of Western Ontario and took his interest in organic photochemistry with him. The Canadians were so impressed that they eventually created a special photochemistry unit, which has functioned to this day. Paul had no difficulty in getting into the British Royal Society even though for those who live in Canada entry is more difficult than for those who live in the United Kingdom. Paul is now retired.

A. Ian Scott

Ian Scott took his doctorate in Glasgow with Ralph Raphael and Sir James Cook. After military service, Ian came to me as a post-doc for about three years, first in London, then in Glasgow. When I left Glasgow, I arranged for Ian (and Paul deMayo) to be appointed lecturers at Imperial College. Ralph Raphael was appointed to succeed me in Glasgow. He also offered Ian a lectureship. I quite understood that Ian would prefer to stay in Glasgow, so he held his lectureship at Imperial College for one month, before returning to Glasgow. There, he continued the work on phenolate coupling that he had begun with me. He completed a brilliant synthesis of griseofulvin, an important mold product, by an ingenious biomimetic route.

Ian Scott with Lord Todd, 1975.

I always underestimated Ian's capacity for mobility. He did so well in phenolate coupling that within a few years he was offered and accepted a professorship at the University of British Columbia. There, his interest in natural products chemistry developed. Again, within a few years, he accepted an offer to become a professor in one of the founding chairs of chemistry in the New University of Sussex in the United Kingdom. He accepted this and started his well-known work on penicillin and other antibiotics. I thought that after six different academic appointments, Ian would stay in the United Kingdom, but this was not to be. After some years, he moved to Yale University for the next decade. Texas A&M then made him one of those irresistible offers, and so he moved to College Station. But the saga had not ended. After a few years, the prestigious chair of organic chemistry at the University of Edinburgh became vacant. Ian was tempted to return to his roots. His application was received with enthusiasm by the British establishment and the appointment was made. The return of the exile did not last long; before the year was out, Ian was back at Texas A&M.

What happened was not the fault of the University of Edinburgh, working within the limits of the British system, but the irresistible pull of life in Texas. A person changes over the years in a different country, even if he does not realize it. Also, the family life of children and grandchildren cannot be transplanted easily from one side of the Atlantic to the other. Commendably, Ian's wife has always supported and helped him in all his moves.

Be these considerations as they may, Ian quickly reestablished fundamental scientific work at College Station. His long-running program on the biosynthesis of Vitamin B_{12} is approaching a climax. Combining more chemical biosynthetic studies with molecular biology is giving spectacular results. Soon, each step in the very complex biosynthesis will be identified in terms of its unique enzyme, available in quantity by cloning the appropriate gene.

Ian's other main interest is in the mechanisms of enzyme action, as studied by high-field NMR spectroscopy and using all the sophisticated techniques now available. In principle, NMR spectroscopic techniques are better than X-ray crystallography because they permit the enzyme and substrate interaction to be studied in its natural medium. This is an important field for the end of this century and the beginning of the next.

My personal relations with Ian have always been cordial, and I have supported him as best I could in his peripatetic career. I think that I understand how emotion can sometimes prevail over reason. It was, therefore, with pleasure that I learned of the feasibility of joining him as a professor at Texas A&M. The instrumentation there is excellent, and the work ethic of the University is above reproach. I have the impression that they appreciate my own habits of beginning the working day at 3–4 in the morning and finishing about 7 in the evening. The rest of my time is for Morpheus (about 4 hours) and my wife (about 4 hours).

Ian and I share a common passion for organic chemistry, whether it be expressed in biosynthesis and molecular biology of enzymes (difficult) or in the invention of new reactions (easy). Long may it endure!

References

1. (a) Barton, D. H. R. *J. Chem. Soc.* **1949**, 148; (b) Barton, D. H. R.; Mugdan, M. *J. Soc. Chem. Ind. London* **1950**, *69*, 75.

2. Barton, D. H. R.; Howlett, K. E. *J. Chem. Soc.* **1949**, 155.

3. Ibid. **1949**, 165.

4. Barton, D. H. R.; Onyon, P. F. *Trans. Faraday Soc.* **1949**, *45*, 725.

5. (a) Barton, D. H. R.; Head, A. J. ibid. **1950**, *46*, 114; (b) Barton, D. H. R.; Howlett, K. E. *J. Chem. Soc.* **1951**, 2033; (c) Barton, D. H. R.; Head, A. F.; Williams, R. J. *J. Chem. Soc.* **1951**, 2039.

6. Barton, D. H. R.; Onyon, P. F. *J. Am. Chem. Soc.* **1950**, *72*, 988.

7. Alexander, P.; Barton, D. H. R. *Biochem. J.* **1943**, *37*, 463.

8. Barton, D. H. R.; Jones, E. R. H. *J. Chem. Soc.* **1944**, 659.

9. Freudenberg, K. *Ber. Dtsch. Chem. Ges.* **1933**, *66*, 177.

10. Bernstein, S.; Kauzmann, W. J.; Wallis, E. S. *J. Org. Chem.* **1942**, *7*, 103.

11. (a) Barton, D. H. R. *J. Chem. Soc.* **1945**, 813; (b) ibid. **1946**, 512; (c) ibid. **1946**, 1116.

12. (a) Barton, D. H. R.; Cox, J. D. *J. Chem Soc.* **1948,** 1354; (b) ibid. **1949,** 214; (c) ibid. **1949,** 219. See also (d) Barton, D. H. R.; Miller, E. ibid. **1949,** 337; (e) Barton, D. H. R.; Cox, J. D.; Holness, N. J. ibid. **1949,** 1771.

13. Barton, D. H. R.; Cox, J. D. ibid. **1948,** 1354.

14. For example, Barton, D. H. R.; Gökturk, A. K.; Jankowski, K. *J. Chem. Soc., Perkin Trans.* **1985,** 2109.

15. (a) Barton, D. H. R.; Klyne, W. *Chem. Ind.* **1948,** 755. See also (b) Barton, D. H. R. *Angew. Chem.* **1949,** *61,* 57.

16. Plattner, P. A.; Heusser, H.; Troxler, F.; Segre, A. *Helv. Chim. Acta* **1948,** *31,* 852.

17. Barton, D. H. R.; Rosenfelder, W. J. *Helv. Chim. Acta* **1949,** *32,* 975.

18. Barton, D. H. R. *Experientia* **1950,** *6,* 316.

19. (a) Barton, D. H. R.; Rosenfelder, W. J. *Nature (London)* **1949,** *164,* 316; (b) idem *J. Chem. Soc.* **1949,** 2459.

20. Barton, D. H. R. *J. Chem. Soc.* **1949,** 2174.

21. Barton, D. H. R.; Head, A. J.; Williams, R. J. *J. Chem. Soc.* **1952,** 453.

22. Simonsen, Sir John; Barton, D. H. R. In *The Terpenes;* Cambridge University: Cambridge, 1952; Vol. 3.

23. Woodward, R. B.; Brutschy, F. J.; Baer, H. *J. Am. Chem. Soc.* **1948,** *70,* 4216.

24. Woodward, R. B.; Sondheimer, F.; Taub, D.; Heusler, K.; McLamore, W. M. *J. Am. Chem. Soc.* **1951,** *73,* 2403.

25. Woodward, R. B.; Sondheimer, F.; Taub, D. *J. Am. Chem. Soc.* **1951,** *73,* 3547.

26. Ibid. **1951,** *73,* 3548.

27. Ibid. **1951,** *73,* 4057.

28. (a) Cardwell, H. M. E.; Cornforth, J. W.; Duff, S. R.; Holtermann, H.; Robinson, R. *Chem. Ind.* **1951,** 389; (b) idem *J. Chem. Soc.* **1953,** 361.

29. Barton, D. H. R. In *Stereochemistry of Organic and Bioorganic Transformations;* Bartmann, W.; Sharpless, K. B., Eds.; VCH Verlagsgesellschaft: Weinheim, 1987.

30. Ramsay, O. B. In *Stereochemistry*; Heyden: London, 1981.

31. Bastiansen, O.; Hassel, O. *Nature (London)* **1946**, *157*, 765.

32. Kemp, J. D.; Pitzer, K. S. *J. Chem. Phys.* **1936**, *4*, 749; (b) idem *J. Am. Chem. Soc.* **1937**, *50*, 276; (c) Pitzer, K. S. *J. Chem. Phys.* **1937**, *5*, 469, 473, 752.

33. Corin, E.; Walter, J.; Eyring, H. *J. Am. Chem. Soc.* **1939**, *61*, 1876.

34. Langeth, A.; Bak, B. *J. Chem. Phys.* **1940**, *8*, 403.

35. Hassel, O. *Tidsskr. Kjemi Bergves. Metall.* **1943**, *3*, 32, translated into English in *Top. Stereochem.* **1971**, 6.

36. Dostrovsky, I.; Hughes, E. D.; Ingold, C. K. *J. Chem. Soc.* **1946**, 173.

37. Westheimer, F. H.; Mayer, J. E. *J. Chem. Phys.* **1946**, *14*, 733.

38. Barton, D. H. R. *J. Chem. Soc.* **1948**, 1197.

39. Barton, D. H. R.; Schmeidler, G. A. *J. Chem. Soc.* **1948**, 1197.

40. Fieser, L. F. *Experientia* **1950**, *6*, 312.

41. Birch, A. J. *Ann. Rep. Prog. Chem.* **1951**, *48*, 192.

42. Fieser, L. F.; Fieser, M. In *Natural Products Related to Phenanthrene*; Reinhold: New York, 1949.

43. Barton, D. H. R. *J. Chem. Soc.* **1953**, 1027.

44. Eliel, E. L.; Allinger, N. L.; Angyal, S. J.; Morrison, G. A. In *Conformational Analysis*; Wiley: New York, 1965.

45. Seeman, J. I. *Chem. Rev.* **1983**, *83*, 83.

46. Abd El Rahim, A. M.; Carlisle, C. H. *Chem. Ind.* **1954**, 279.

47. Barnes, C. S.; Barton, D. H. R.; Fawcett, J. C.; Thomas, B. R. *J. Chem. Soc.* **1953**, 576, and references there cited.

48. Barton, D. H. R.; Ives, D. A. J.; Kelly, R. B.; Woodward, R. B.; Patchett, A. A. *Chem. Ind.* **1954**, 605.

49. Woodward, R. B.; Patchett, A. A.; Barton, D. H. R.; Ives, D. A. J.; Kelly, R. B. *J. Am. Chem. Soc.* **1954**, *76*, 2852.

50. Todd, Lord; Cornforth, Sir J. *Biogr. Mem. Fellows R. Soc.* **1981**, 629.

51. Woodward, R. B.; Patchett, A. A.; Barton, D. H. R.; Ives, D. A. J.; Kelly, R. B. *J. Chem. Soc.* **1957**, 1131.

52. Barton, D. H. R. *J. Chem. Soc.* **1951**, 1444.

53. Robertson, J. M.; Todd, G. *J. Chem. Soc.* **1955**, 1254.

54. (a) Barton, D. H. R.; Lindsey, A. S. *Chem. Ind.* **1951**, 313; (b) idem *J. Chem. Soc.* **1951**, 2988.

55. (a) Barton, D. H. R.; Bruun, T.; Lindsey, A. S. *Chem. Ind.* **1951**, 910; (b) idem *J. Chem. Soc.* **1952**, 2210.

56. Aebi, A.; Barton, D. H. R.; Burgstahler, A. W.; Lindsey, A. S. *J. Chem. Soc.* **1954**, 4659.

57. See Barton, D. H. R.; Cheung, H. T.; Cross, A. D.; Jackman, L. M.; Martin-Smith, M. *J. Chem. Soc.* **1961**, 5061.

58. See Barton, D. H. R.; Pradhan, S. K.; Sternhell, S.; Templeton, J. F. *J. Chem. Soc.* **1961**, 255.

59. See Barton, D. H. R.; Sutherland, J. K. *J. Chem. Soc.* **1965**, 1769.

60. Barton, D. H. R.; Lewis, D. A.; McGhie, J. F. *J. Chem. Soc.* **1957**, 2907.

61. Barton, D. H. R.; Head, A. J.; May, P. J. *J. Chem. Soc.* **1957**, 935.

62. Barton, D. H. R.; McCapra, F.; May, P. J.; Thudium, F. *J. Chem. Soc.* **1960**, 1297.

63. Allinger, N. L.; Lane, G. A. *J. Am. Chem. Soc.* **1974**, *96*, 2937.

64. Barton, D. H. R. *Chem. Br.* **1973**, *9*, 149.

65. Pummerer, R.; Puttfarcken, H.; Schopflocken, P. *Ber. Dtsch. Chem. Ges.* **1925**, *58*, 1808.

66. Barton, D. H. R.; Deflorin, A. M.; Edwards, O. E. *Chem. Ind.* **1955**, 1039.

67. Idem *J. Chem. Soc.* **1956**, 530.

68. Barton, D. H. R.; Cohen, T. *Festschr. Prof. Dr. Arthur Stoll Siebzigsten Geburtstag* **1957**, 117.

69. Haynes, L. J.; Stuart, K. L.; Barton, D. H. R.; Kirby, G. W. *Proc. Chem. Soc. London* **1963**, 280; (b) ibid. **1964**, 261; (c) idem *J. Chem. Soc. C* **1966**, 595.

70. Summarized in Barton, D. H. R.; Kirby, G. W.; Steglich, W.; Thomas, G. M.; Battersby, A. R.; Dobson, T. A.; Ramuz, H. *J. Chem. Soc.* **1965**, 2423.

71. Barton, D. H. R.; Kirby, G. W.; Steglich, W.; Thomas, G. M. *Proc. Chem. Soc. London* **1963**, 203.

72. Barton, D. H. R.; Bracho, R. D.; Potter, C. J.; Widdowson, D. A. *J. Chem. Soc., Perkin Trans. 1* **1974**, 2278.

73. Unpublished observations carried out in collaboration with Edward Wheeler, 1954.

74. Barton, D. H. R.; De Mayo, P.; Shafiq, M. *J. Chem. Soc.* **1957**, 929.

75. Idem *Proc. Chem. Soc. London* **1957**, 205.

76. Idem *J. Chem. Soc.* **1958**, 140.

77. Arigoni, D.; Bosshard, H.; Bruderer, H.; Büchi, G.; Jeger, O.; Krebaum, L. *J. Helv. Chim. Acta* **1957**, *40*, 1732.

78. Cocher, W.; Crowley, K.; Edward, J. T.; McMurry, T. B. H., Stuart, E. R. *J. Chem. Soc.* **1961**, 1215.

79. Barton, D. H. R.; McGhie, J. F.; Rosenberger, M. *J. Chem. Soc.* **1961**, 1215.

80. Barton, D. H. R.; Quinkert, G. *Proc. Chem. Soc. London* **1958**, 197.

81. Idem *J. Chem. Soc.* **1960**, 1.

82. Schneider, K. R.; Dürmer, G.; Hache, K.; Stegk, A.; Quinkert, G.; Barton, D. H. R. *Chem. Ber.* **1977**, *110*, 3582.

83. Akhtar, M.; Barton, D. H. R.; Beaton, J. M.; Hartmann, A. G. *J. Am. Chem. Soc.* **1963**, *83*, 1512.

84. (a) Barton, D. H. R.; Beaton, J. M.; Geller, L. E.; Pechet, M. M. *J. Am. Chem. Soc.* **1960**, *82*, 2640; (b) Barton, D. H. R.; Beaton, J. M. *J. Am. Chem. Soc.* **1960**, *82*, 2641; (c) ibid. **1961**, *83*, 4083.

85. Barton, D. H. R.; Basu, N. K.; Day, M. J.; Hesse, R. H.; Pechet, M. M.; Starrat, A. N. *J. Chem. Soc., Perkin Trans. 1* **1975**, 2243.

86. Barton, D. H. R.; Crich, D.; Kretzschmar, G. *J. Chem. Soc., Perkin Trans. 1* **1986**, 39.

87. Barton, D. H. R.; Prabhakar, S. K. *J. Chem. Soc., Perkin Trans. 1* **1974**, 781.

88. Barton, D. H. R.; Miller, E. *J. Am. Chem. Soc.* **1950**, *72*, 1066.

89. Barton, D. H. R.; King, J. F. *J. Chem. Soc.* **1958**, 4398.

90. Achmatowicz, S.; Barton, D. H. R.; Magnus, P. D.; Poulton, G. A.; West, P. J. *J. Chem. Soc., Chem. Commun.* **1971**, 1014.

91. Barton, D. H. R.; Bolton, M.; Magnus, P. D.; West, P. J.; Porter, G.; Wirz, J. *J. Chem. Soc., Chem. Commun.* **1972**, 632.

92. Barton, D. H. R.; Bolton, M.; Magnus, P. D.; West, P. J. *J. Chem. Soc., Perkin Trans. 1* **1973**, 1580, and earlier papers.

93. Barton, D. H. R.; McCombie, S. W. *J. Chem. Soc., Perkin Trans. 1* **1975**, 1574.

94. Hartwig, W. *Tetrahedron* **1983**, *39*, 2609.

95. Robins, M. J.; Wilson, J. S.; Hansske, F. *J. Am. Chem. Soc.* **1983**, *105*, 4059.

96. Barton, D. H. R.; Bringmann, G.; Lamotte, G.; Motherwell, W. B.; Hay Motherwell, R. S.; Porter, A. E. A. *J. Chem. Soc., Perkin Trans 1* **1980**, 2657; Barton, D. H. R.; Bringmann, G.; Motherwell, W. B. *J. Chem. Soc., Perkin Trans 1* **1980**, 2665.

97. Barton, D. H. R.; Crich, D.; Motherwell, W. B. *J. Chem. Soc., Chem. Commun.* **1985**, *41*, 3901.

98. Barton, D. H. R.; Zard, S. Z. *Philos. Trans. R. Soc. London B* **1985**, *311*, 505; idem *Pure Appl. Chem.* **1986**, *58*, 675.

99. Ingold, K. U.; Lusztyk, J.; Maillard, B.; Walton, J. C. *Tetrahedron Lett.* **1988**, *29*, 917.

100. Barton, D. H. R.; George, M. V.; Tomoeda, M. *J. Chem. Soc.* **1962**, 1967.

101. Delduc, P.; Tailhan, C.; Zard, S. Z. *J. Chem. Soc., Chem. Commun.* **1988**, 308.

102. *See* Barton, D. H. R.; Gastiger, M. J.; Motherwell, W. B. *J. Chem. Soc., Chem. Commun.* **1983**, 41.

103. Tabushi, I.; Nakajima, T.; Seto, K. *Tetrahedron Lett.* **1980**, *21*, 2565.

104. Barton, D. H. R.; Boivin, J.; Ozbalik, N.; Schwartzentruber, K. M.; Jankowski, K. *Tetrahedron Lett.* **1985**, *26*, 447.

105. Bower, B. K.; Tennent, H. G. *J. Am. Chem. Soc.* **1972**, *94*, 2512.

106. Barton, D. H. R.; Crich, D.; Motherwell, W. B. *Tetrahedron* **1985**, *41*, 3901; Barton, D. H. R.; Zard, S. Z. *Pure Appl. Chem.* **1986**, *58*, 675.

107. Balavoine, G.; Barton, D. H. R.; Boivin, J.; Gref, A.; Ozbalik, N.; Rivière, H. *Tetrahedron Lett.* **1986**, *27*, 2849; idem *J. Chem. Soc., Chem. Commun.* **1986**, 1727.

108. Barton, D. H. R.; Boivin, J.; Le Coupanec, P. *J. Chem. Soc., Chem. Commun.* **1987**, 1379.

109. Walling, C. *Acc. Chem. Res.* **1975**, *8*, 125.

110. For reviews on Cp-450, see (a) Guengerich, F. P.; Macdonald, T. L. *Acc. Chem. Res.* **1984**, *17*, 9; (b) White, R. E.; Coon, M. J. *Ann. Rev. Biochem.* **1980**, *19*, 315; (c) Ullrich, V. *Top. Curr. Chem.* **1979**, *83*, 67; (d) Groves, J. T. *Adv. Inorg. Biochem.* **1979**, 119.

111. (a) Groves, J. T.; Krishnan, S.; Avaria, G. E.; Nemo, T. E. In *Biomimetic Chemistry*; Dolphin, D.; McKenna, C.; Murakami, Y.; Tabushi, I., Eds.; Advances in Chemistry 191; American Chemical Society: Washington, DC, 1980; p 277; (b) Tabushi, I.; Koga, N. In *Biomimetic Chemistry*; Dolphin, D.; McKenna, C.; Murakami, Y.; Tabushi, I., Eds.; Advances in Chemistry 191; American Chemical Society: Washington, DC, 1980; p 291; (c) Nee, M. W.; Bruice, T. C. *J. Am. Chem. Soc.* **1982**, *104*, 6123; (d) Lindsay Smith, J. R.; Sleath, P. R. *J. Chem. Soc., Perkin Trans. 2* **1982**, 1009; (e) Lee, W. A.; Bruice, T. C. *J. Am. Chem. Soc.* **1985**, *107*, 513.

112. (a) Hayaishi, O.; Katagiri, O.; Rothberg, S. *J. Am. Chem. Soc.* **1955**, *77*, 5450; (b) Mason, H. S.; Foulks, W. L.; Paterson, E. *J. Am. Chem. Soc.* **1955**, *77*, 2914; for recent reviews on the selective oxidation of hydrocarbons, see Crabtree, R. H. *Chem. Rev.* **1985**, *85*, 245; Green, M. L. H.; O'Hare, D. *Pure Appl. Chem.* **1985**, *57*, 1897; Halpern, J. *Inorg. Chim. Acta* **1985**, *100*, 41; Rothwell, I. P. *Polyhedron* **1985**, *4*, 177.

113. Baldwin, J. E.; Adlington, R. M.; Flitsch, S. L.; Ting, H.-H.; Turner, N. J. *J. Chem. Soc., Chem. Commun.* **1986**, 1305.

114. Corey, E. J.; Nagata, R. *Tetrahedron Lett.* **1987**, *28*, 5391.

115. (a) Shorter, J. *Natural Product Reports* **1987**, *4*, 61; (b) Saltzman, M. D. *Natural Product Reports* **1987**, *4*, 53; (c) Williams, T. *Endeavor*, **1986**, *10*, 107.

Index

Production: Peggy D. Smith
Copyediting and Indexing: A. Maureen Rouhi
Acquisition: Robin Giroux

Books printed and bound by Maple Press, York, PA

Paper meets minimum requirements of American National Standard
for Information Sciences—Permanence of Paper for Printed Library
Materials, ANSI Z39.48–1984 ∞